TERMINAL SEDATION: EUTHANASIA IN DISGUISE?

Terminal Sedation: Euthanasia in Disguise?

Edited by

Torbjörn Tännsjö
Department of Philosophy,
Stockholm University, Sweden

KLUWER ACADEMIC PUBLISHERS
DORDRECHT / BOSTON / LONDON

A C.I.P. Catalogue record for this book is available from the Library of Congress.

ISBN 1-4020-2123-2 (HB)
ISBN 1-4020-2124-0 (e-book)

Published by Kluwer Academic Publishers,
P.O. Box 17, 3300 AA Dordrecht, The Netherlands.

Sold and distributed in North, Central and South America
by Kluwer Academic Publishers,
101 Philip Drive, Norwell, MA 02061, U.S.A.

In all other countries, sold and distributed
by Kluwer Academic Publishers,
P.O. Box 322, 3300 AH Dordrecht, The Netherlands.

Printed on acid-free paper

Printed in the Netherlands.

ABSTRACT

Around 1990 a discussion started concerning a measure within palliative care that had earlier attracted little attention: the sedation of dying patients. The intention behind this anthology is to continue and deepen the discussion. Those who have contributed constitute a mixture of distinguished bioethicists and doctors and nurses with experience of terminal sedation, some of them in favour of, and others in strong opposition to, voluntary euthanasia. They give their view on terminal sedation by answering questions such as: 'Is terminal sedation euthanasia in disguise?', 'Should terminal sedation be a part of standard palliative care?', 'Should terminal sedation be provided at the patient's request?' This is the first book devoted exclusively to this subject. It does not give the final verdict, but it does contain strong defences by very competent thinkers of the most important and influential positions. It has been published in the hope that it will provoke further thought and discussion.

TABLE OF CONTENTS

ACKNOWLEDGEMENTS

Most of the authors of this book were summoned to a conference in Gothenburg, Sweden, in January 2002, to discuss problems to do with terminal sedation. The invited authors have been carefully selected because of their excellence in the field and their highly varied points of view. The conference was organised by the Ethical Board of the Sahlgrenska Academy together with the Department of Philosophy at Gothenburg University. The conference was kindly subsidised by the Swedish Academy of Letters. The conference was chaired with skill and diplomacy by Göran Hermerén, a member of the Academy of Letters and Professor of Medical Ethics at the University of Lund. I thank him and all these institutions for their generous support. A grant from the Swedish Research Council has rendered possible the publishing of the book. I also thank the editors of the *Journal of the American Medical Association* for their kind permission to allow me to reprint an article by Timothy E. Quill, Bernard Lo, and Dan W. Brock (Chapter 1 of this book), introducing some of the key notions in the debate. The constructive advice from two anonymous reviewers for Kluwer was most helpful. And, finally, the patient assistance I received from Jacqueline Bergsma and Nellie Harrewijn, editors at Kluwer, has proved invaluable.

Torbjörn Tännsjö, December, 2003

THE CONTRIBUTORS

Magna Andreen Sachs is a Medical Doctor and Associate Professor at Karolinska Institute, specialised in anaesthesiology. She is an adviser of the Swedish National Board of Health and Welfare.

Dan W. Brock is Professor of Philosophy Emeritus at Brown University. He is now Senior Scientist in the Department of Clinical Bioethics, National Institutes of Health in Washington. He has published extensively in medical ethics. His most recent book, which he has written together with several other distinguished bioethicists, is *From Chance to Choice: Genetics and Justice* (2000).

Daniel Callahan is Director of International Programs for the Hastings Center. He was a co-founder of the Center in 1969. He has published extensively in medical ethics and his most recent books are *False Hopes: Why America's Quest for Perfect Health is a Recipe for failure* (1998), *Promoting Healthy Behavior: How Much Freedom? Whose Responsibility?* (2000) and *What Price Better Health? Hazards of the Research Imperative* (2003).

Johannes JM van Delden is Professor of Medical Ethics at the University Medical Center, Utrecht, the Netherlands. He is also a nursing home physician in the Rosendael nursing home in Utrecht.

Gunnar Eckerdal is a Medical Doctor specialised in palliative care and working at Bräcke Hospice, Gothenburg. He has in some cases refused to terminally sedate his patients at their request, but he has also sedated one of his patients on the presumption that this was what the patient would have asked for, had she been able to communicate with him.

Luke Gormally is a Senior Research Fellow of The Linacre Centre for Healthcare Ethics, London, UK, of which he was Director 1981-2000, and a Research Professor at Ave Maria School of Law, Ann Arbor, Michigan, USA. He is a member of the Pontifical Academy for Life. His most recent books are: Luke Gormally (ed) *Culture of Life - Culture of Death* (2002), Anthony Fisher and Luke Gormally (eds) *Healthcare Allocation: An Ethical Framework for Public Policy* (2001), and Luke Gormally (ed) *Issues for a Catholic Bioethic* (1999).

Helga Kuhse is Senior Honorary Research Fellow at Monash Centre for Human Bioethics. She has published extensively in medical ethics, and her most recent books are *Caring: Nurses, Women, and Ethics* (1997) and, together with Peter Singer, *Unsanctifying Human Life* (2002).

Bernard Lo, MD, is the Director of the CAPS Ethics Core. He is Professor of Medicine and Director of the Program in Medical Ethics at UCSF. His book, *Resolving Ethical Dilemmas: A Guide for Clinicians*, contains chapters on issues arising in the care of HIV-infected patients, including confidentiality and partner notification, decisions about life-sustaining interventions, assisted suicide, patient requests for interventions, refusal by physicians to care for patients, and transmission of HIV in health care settings. Dr. Lo is a member of the Board of Health Sciences Policy of the Institute of Medicine of the Board of Directors of the American Society of Law, Medicine, and Ethics.

Timothy Quill, MD., was the plaintiff in the landmark Quill vs. Vacco Supreme Court decision in 1997. He is Professor of Medicine, Psychiatry and Medical Humanities at the University of Rochester, a co-director of an inpatient Palliative Care Unit, and a primary care internist. He is the author of the books: *Death With Dignity*, *A Midwife Through the Dying Process*, and *Caring for Patients at the End of Life: Facing an Uncertain Future Together*.

Torbjörn Tännsjö is Professor of Practical Philosophy at Stockholm University. He has published extensively in moral philosophy, political philosophy and medical ethics. His most recent books are *Hedonistic Utilitarianism* (1998), *Coercive Care* (1999), and *Understanding Ethics: An Introduction to Moral Theory* (2002).

Simon Woods is a Senior Lecturer at the Policy, Ethics and Lifes Sciences Research Institute (PEALS), University of Newcastle (UK). He is a Fellow of the Institute of Medicine Law and Bioethics at the University of Manchester, is qualified as a nurse and holds bachelor and doctoral degrees in philosophy. His clinical career as a nurse was in the field of oncology and he has conducted empirical and conceptual research within oncology and palliative care and has published widely in medical ethics.

INTRODUCTION

TERMINAL SEDATION DURING THE 1990s

During the 1990s a discussion took place in scholarly journals concerning a measure within palliative care that had earlier attracted little attention, to wit, the sedation of dying patients. There seem to have been two main reasons why the practice came under debate. On the one hand, some people felt that, when palliative medicine had advanced and methods to control symptoms had improved, it was no longer justified to sedate the patients in a manner that had often been done in the past. The system of terminal sedation had turned into 'euthanasia in disguise' or 'slow euthanasia'.[1] On the other hand, there were people sympathetic to the recently established Dutch system of euthanasia, people who agreed that terminal sedation was euthanasia in disguise, but who felt that, if it is not objectionable to *sedate* dying patients at their request, then why should it not be permitted for doctors to *kill* dying patients at request?[2] From these two motives a discussion about terminal sedation gained momentum. The intention behind this anthology is to continue and deepen this discussion.

The anthology starts off with a chapter where an influential article from the 1990s has been reprinted. The article was written by Timothy Quill, Bernard Lo, and Dan Brock, who do not see terminal sedation as morally problematic as such, or at least as no *more* problematic than the forgoing of life-saving treatment (which *is*, according to these authors, problematic indeed, not *per se*, but because it lends itself to possible abuse). In this article some notions (other than the notion of terminal sedation) of importance to the discussion of end-of-life decisions are introduced, defined and discussed, to wit, the notions of a patient voluntarily stopping eating and drinking (VSED), physician-assisted suicide (PAS), and voluntary active euthanasia (VAE). This article played a crucial role in the scholarly discussion and also in my own involvement in the debate.

My own interest in terminal sedation was raised when, in 1998, as a member of the Ethics Committee of the Swedish National Board of Health and Welfare, I was asked to make up my mind about three cases where terminal patients had been definitely put to sleep. This had happened in 1996 and 1997 in a Swedish Hospice (Kålltorp). In order to grant his patients a peaceful death their physician, Dr. Mats Holmberg, had terminally sedated them. In two of the three cases Dr. Holmberg had

Torbjörn Tännsjö (ed.), Terminal Sedation: Euthanasia in Disguise?, xiii-xxiii.

acted on the patient's explicit request; in one of the cases, however, he had acted on the mere presumption that terminal sedation was what the patient would have requested, had he still been able to communicate with his doctor. I thought Dr. Holmberg had done the right thing in all of the cases, but there was no unanimity about this in the Ethics Committee.

I began to read about terminal sedation and I became aware of the discussion that had been going on in scholarly journals during the 1990s. There were other cases described there. One of them was a much publicised Australian one, also commented upon by Helga Kuhse in Chapter 6 of this book. When the Rights of the Terminally Ill Act, permitting voluntary euthanasia in the Northern Territory, was invalidated in 1997 by Federal Legislation, one terminally ill patient, Ester Wild, had fulfilled all the requirements laid down in the Act, but her doctor, Philip Nitschke, was no longer able to end her life by euthanasia. Instead, and with the patient's consent, he provided terminal sedation and Ester Wild died four days later, in a state of medically induced unconsciousness.

In the ethics committee of the Swedish National Board of Health and Welfare the opinion was divided. Some thought that Dr. Mats Holmberg had made the wrong decision in all of the three cases presented to us. Patients should never be terminally sedated, they claimed. Others argued that while it was correct terminally to sedate the two patients who had made an explicit request for terminal sedation, it was wrong to sedate the third patient on the mere presumption that this was what he would have asked for, had he been communicable. Yet others argued that only the patient who was not communicable should have been terminally sedated. This kind of treatment, they claimed, should never be offered to a competent patient (who has conscious life left to lose, and who should therefore always be given a chance to change his or her mind).

This diversity of opinions was perhaps only to be expected. For it surfaced from the discussion in the scholarly journals that terminal sedation was indeed a most controversial issue. Some tend to argue that terminal sedation should never take place. Others argue, on the contrary, that it should be a standard option within good palliative care. Some argue that terminal sedation is euthanasia in disguise, and should for this sake be prohibited. Some agree that it is indeed euthanasia in disguise (or 'slow euthanasia') but argue that, for all that, it should (together with voluntary euthanasia) be permitted.

The Swedish National Board of Health and Welfare tried to get the doctor who had sedated his patients, Dr. Mats Holmberg, convicted of murder or manslaughter (in spite of the fact that the opinion in the ethics committee was divided). A similar case soon appeared in Norway, where at Baerum Hospital a Dr. Stig Ottesen was accused (by a Swedish colleague working in the same hospital) of having resorted to illegal euthanasia when he had terminally sedated some of his patients. Several experts on palliative medicine, from both Sweden and Norway, were called upon to give their testimony, and the Swedish medical experts were once again unanimous in their verdict: terminal sedation is euthanasia in disguise and, hence, illegal. Both the Swedish and the Norwegian cases were much publicised.

It is noteworthy that both Dr. Philip Nitschke, when he resorted to terminal sedation of Ester Wild, and Dr. Mats Holmberg, when he sedated two of his patients, provided terminal sedation to patients who had asked for euthanasia. Dr. Philip

Nitschke did not hesitate to provide voluntary euthanasia but, in the circumstances he was prohibited by the law to do so. The law also prohibited Dr. Holmberg from providing voluntary euthanasia but, in contradistinction to Dr. Philip Nitschke, Dr. Holmberg had no inclination to provide voluntary euthanasia. In fact, Dr. Holmberg is a determined opponent of voluntary euthanasia. And yet, in the circumstances, these two doctors reached the same decision. They both sedated their patients.

This observation gave me the idea that terminal sedation might be a possible compromise position in countries where the discussion about voluntary euthanasia has reached a deadlock. In order to be a viable option this compromise requires, of course, that terminal sedation is not euthanasia in disguise. This was the belief of Dr. Mats Holmberg and in this belief of his I was strongly inclined to concur, in spite of many medical experts of a contrary opinion.

At the world congress of the International Association of Bioethics in Tokyo, in the fall of 1998, the very same year that the Kålltorp case had been brought forward to the Ethics Committee of the Swedish National Board of Health and Welfare, I talked to my friends and colleagues Professor Helga Kuhse and Professor Dan Brock about terminal sedation (in a discussion following upon a session on euthanasia). I share their favourable view of voluntary euthanasia but, obviously, in many countries euthanasia is no option. So I put forward my proposal that, in countries where euthanasia is not an option, terminal sedation might constitute a reasonable compromise. I spotted no sign of sympathy with my suggestion, however. These distinguished bioethicists seemed both to hold the opinion that terminal sedation is indeed euthanasia in disguise so, as a matter of consistency, voluntary euthanasia and terminal sedation stand and fall together. If one is legalised, then so should the other.

At the next world congress of the International Association of Bioethics, which took place in London, in the Fall of 2000, I gave a talk on terminal sedation, which has subsequently been published.[3] I had now worked out my argument to the effect that terminal sedation is not euthanasia in disguise and I put forward a more elaborate version of the idea that terminal sedation might constitute a reasonable compromise in the discussion about voluntary euthanasia, which has, in many countries, reached a deadlock. Professor Helga Kuhse was chairing the session, and I realised that I had not convinced her (or Professor Dan Brock, with whom I corresponded on the question). I therefore decided to arrange an international symposium, where the problem of terminal sedation was thoroughly penetrated and where my proposal was scrutinized. The time was ripe for a new intervention into the problem of terminal sedation.

I was extremely lucky when it turned out that all the people I wanted to participate in the symposium were able to do so. In particular, I had hoped to be able to gather a mixture of internationally distinguished bioethicists, both those who advocate, and those who oppose, euthanasia, as well as a skilled physician, who had his own experience of terminal sedation, and at least one of the medical experts who had claimed in relation to both the Swedish and the Norwegian cases that terminal sedation is euthanasia in disguise and therefore illegal. All these desiderata were fulfilled. I was also pleased to note that many of the people who had been directly involved in the Scandinavian cases were present at the conference, such as the Baerum doctor who had been accused of having terminally sedated his patients (Dr.

Stig Ottesen), together with the Swedish doctor who had 'blown the whistle' at Baerum hospital (Dr. Carl Magnus Edenbrant, now chairing the Swedish Association of Palliative Medicine).

I think it fair to say that during the symposium, we were all capable of having a cool and rational discussion about the hotly contested issue of terminal sedation. This does not mean, however, that any consensus was arrived at.

Let me now introduce those who contributed to the discussion as invited speakers and their respective chapters in this anthology.

2. THE CONTRIBUTIONS

In the next chapter we are given a hint of the state of the art. One of the most influential contributions to the discussion during the 1990s is here reprinted. In the following Chapter 2, I try to set the stage for a deepened discussion by introducing the three Swedish cases. This gives me the opportunity to define terminal sedation. It has transpired that not only the practice of terminal sedation, but the notion as well is controversial. There arc many possible ways of defining the notion of terminal sedation. As the notion is understood in the present context, however, it is crucial that the patient is put *definitely* into a state of *unconsciousness* (supposed to go on until the patient is dead) while, at the same time, artificial nutrition and hydration of the patient are withheld.

In Chapter 2 I argue against the view put forward by Professor Dan Brock and others in the article reprinted in Chapter 1 that terminal sedation, thus conceived, is euthanasia in disguise (or 'slow euthanasia'). My argument rests on the observation that while the sedation of the patient may mean that the patient is actively killed (by complications related to, and caused by, the sedation), the death of the patient is not, in that case, intended by the doctor, but merely foreseen. So this is different from euthanasia. As for the withdrawal of artificial nourishment and hydration of the patient, the intention may certainly be to hasten death, I submit. However, since the means of doing so are passive rather than active, this is once again different from euthanasia.

In this chapter I also distinguish between three positions with respect to terminal sedation. (1) That it should never take place, (2) that it could take place, but only as a last resort, and (3) that it should be considered a normal part of palliative care and that a doctor should be free to provide it at the terminally ill patient's request. I defend the third position. In particular I criticise the view put forward by Dr. Balfour Mount that spiritual suffering is special and should not be alleviated by sedation.

In this chapter I also put forward the thesis that terminal sedation should not only be provided at the patient's explicit request but also, sometimes, when the patient is incompetent. When the patient is incompetent, terminal sedation may be provided on the presumption that this is what the patient *would* have asked for *had* he or she *been* able to communicate with the doctor.

The participants in the symposium had all of them rcad my contribution in advance and the participants had been encouraged to react to some, or all, of my theses. None of the contributors had read each other's contributions, however, so there is no room for any dialogue between them in this volume.

Associate Professor Magna Andreen Sachs, who is a medical doctor specialised in anaesthesiology and intensive care at Karolinska Institute, has written Chapter 3. She had been asked to give medical advice in both the Swedish (Kålltorp) and the Norwegian (Baerum) cases. Dr. Andreen Sachs defends the position that terminal sedation (as defined in this chapter) should never be provided. Terminal sedation is an illegal form of euthanasia. Terminal sedation is indeed euthanasia in disguise.

In her contribution Dr. Andreen Sachs observes that there is no unanimity about the definition of the notion of terminal sedation, and she clarifies the terminology in the field. Her clarification of the notion makes it possible for her to render plausible how some people can claim that terminal sedation must never take place (and does hardly ever take place) while others claim that terminal sedation often happens. If 'sedation' is conceived of as *light* sedation, she observes, where the patient is kept communicable, or at least 'wakeable', terminal sedation is not controversial. However, when the goal of the sedation is to render the patient unconscious, i.e., when the patient is sedated into 'oblivion', and where the decision to sedate the patient into oblivion is definitive, i.e., where the sedation of the patient is permanent, and not merely intermittent, *then* we are faced with a practice that is, in Dr. Andreen Sachs' understanding, at variance with standard medical practice. Thus understood, she claims, terminal sedation hardly ever takes place within palliative care.

In Chapter 4. Dr. Gunnar Eckerdal, a distinguished Swedish expert on palliative medicine, with much experience of actually caring for dying patients, presents three actual cases from his own practice. One of these cases is of great importance to the subject matter of this book. It is obvious from Eckerdal's description of the case, that this is 'terminal sedation' in the strong sense of the term defined in this chapter. And yet, in spite of this, Eckerdal defends the decision to sedate his patient.

It should also be noted that this is a case of terminal sedation where no request had been made from the patient, who was incompetent. It is rather a case of non-voluntary terminal sedation where deep and indefinite unconsciousness has been presumed to be in the best interest of both the patient and her close ones. As a matter of fact, this case is very similar to the most controversial case at Kålltorp Hospice (the sedation of Mr. Winter, as described in Chapter 2).

However, Dr. Eckerdal also describes a case where he refused to sedate a terminally ill patient, who had made an explicit request for terminal sedation. In the interest of the autonomy of this patient, Dr. Eckerdal submits, he did not accede to this request.

A competent patient, according to Dr. Eckerdal, should always be given the possibility to reconsider his or her decision. But if the patient is sedated into oblivion, there is no possibility for the patient to change his or her mind.

Crucial to Eckerdal's argument is the assumption that while a patient has an absolute right to refuse treatment, no patient has any corresponding right to request a certain treatment. The doctor, and the doctor alone, must take the full responsibility for the choice of any therapeutic measure. This is also illustrated by Dr. Eckerdal with a more mundane case, where he refused to prescribe antibiotics to a patient who asked for such treatment.

When editing the volume it occurred to me that one medical voice was missing. No nurse had taken part in the conference. So I asked Dr. Simon Woods, who is both a nurse and a philosopher specialised in palliative medicine, to contribute a chapter from a nursing perspective. He has written Chapter 5 of the volume. In this chapter he outlines some special virtues pertaining to nursing in a palliative context, he questions the notion of terminal sedation used in the volume, and suggests, as an alternative, the notion of palliative sedation. In particular, he emphasises, we should not in our definition of terminal sedation (or, rather, in our definition of palliative sedation) include any reference to decisions regarding hydration and nutrition as part of the same clinical decision to utilise sedation. However, even with his interpretation of palliative care, there is room in some cases also for the sedation into oblivion of patients experiencing severe existential pain, and for the withdrawal of artificial hydration and nutrition in these cases — at the dying patient's request. An interesting feature of the chapter is that the question of palliative care is here put into a broad European perspective where references are made to a recent research project (the PALLIUM project), where palliative care has been studied in several European countries.

Chapter 6 is written by Professor Helga Kuhse at Monash Centre for Human Bioethics. Professor Kuhse, being a distinguished advocate of euthanasia at the patient's request, has no objection to terminal sedation, of course, but considers it a form of slow euthanasia. So if we want to accept terminal sedation, she claims, we should not hesitate to accept euthanasia as well. In particular, terminal sedation is at variance with the Sanctity-of-Life Doctrine. However, upon closer examination, there seem to exist good reasons to give up the Sanctity-of-Life Doctrine. Her reasons why we should give up the Sanctity-of-Life Doctrine are well known, of course. But why is terminal sedation inconsistent with the Doctrine?

The main reasons that terminal sedation cannot be accepted by adherents of the Sanctity-of-Life Doctrine, according to Kuhse, are twofold. My argument to the contrary notwithstanding, the sedation of the patient, when, as a matter of fact, no artificial nutrition or hydration will be undertaken, is intentionally and actively to hasten death (to kill the patient). Adherents of the Sanctity-of-Life Doctrine cannot accept this. Furthermore, the decision to forgo artificial nutrition and hydration is as such a decision aiming at the death of the patient. And, according to Professor Kuhse, this is prohibited by the Sanctity-of-Life Doctrine, my argument to the contrary notwithstanding. And this is prohibited by the Sanctity-of-Life Doctrine even if could be described as merely allowing nature to take its course.

It is thus obvious that the editor of the volume, and Professor Kuhse, disagree about how best to understand the doctrine of the Sanctity of Life.

Chapter 7 is written by Professor Dan Brock. Professor Brock has earlier, in the article reprinted in Chapter 1 of this book, written together with Bernard Lo and Timothy Quill, argued that terminal sedation is euthanasia in disguise, or 'slow euthanasia', to use Professor Kuhse's term. In his contribution to this volume he modifies his position somewhat. He is still sceptical about the active/passive distinction but he does not insist any more that the withholding of treatment of the sedated patient means that the doctor intentionally kills the patient. His perspective on the question of terminal sedation is now different. From the point of view of a moral rights theory he argues that there is a moral right to control the time and manner of one's death that includes terminal sedation as well as voluntary euthanasia. But he concedes that that right and presumption might be overridden by sufficiently bad consequences of permitting them, for example if doing so led to many patients being killed who did not wish to die. He notes that the common assumption is that such practices as voluntary euthanasia and physician assisted suicide are more subject to abuse than merely forgoing life-sustaining treatment or pain relief that may hasten death. But since the latter practices are considered part of standard palliative care, they are in fact open to abuse. They are not strictly regulated in the way adherents of voluntary euthanasia want that practice to be regulated. To the extent that terminal sedation is viewed as part of standard medical practice, it will be seen as an appropriate option for choice either by competent patients or by surrogates of incompetent patients, and not as requiring any special safeguards or procedures. This means that while terminal sedation should be permitted, it should be treated in the way voluntary euthanasia and physician assisted suicide is treated – and in the way forgoing life-sustaining treatment *should* be treated.

While editing these chapters it occurred to me that there was one more voice missing at the symposium. A strongly contested question had turned out to be whether the Sanctity-of-Life Doctrine is, or is not, consistent with terminal sedation. But at the symposium no speaker had appeared, who was sympathetic to this doctrine. This looked strange, it seemed to me. I do not say that one must adhere to a certain doctrine in order to be able to give an adequate and fruitful interpretation of it. However, it is utterly strange if one of the most hotly contested issues in a volume like this is whether terminal sedation is consistent with the Sanctity-of-Life Doctrine, while, at the same time, no one adhering to this doctrine has any say whatsoever about this matter. So, in order to correct this mistake of mine, I asked a distinguished adherent of the Sanctity-of-Life Doctrine to contribute a chapter to the volume, in spite of the fact that, through an omission on my part, he had not been present at the symposium. Professor Luke Gormally agreed to contribute a chapter. He is the author of Chapter 8 of this volume. He there states his reasons why, in his view, terminal sedation, if it *aims* at hastening the death of patients by rendering them comatose and depriving them of food and fluids, is inconsistent with the Sanctity-of-Life Doctrine. According to Professor Gormally, terminal sedation, thus conceived, is indeed euthanasia in disguise. Even if such a practice is consistent with standard ethical and legal thinking in the field, this only means that standard ethical and legal thinking is at variance with the Sanctity-of-Life Doctrine. Here Professor Gormally sides with Professor Kuhse in the interpretation of the doctrine. Neither Professor Gormally nor Professor Kuhse, then, is prepared to allow the Sanctity-of-Life Doctrine to be tempered with the use of the active and passive distinction, as has been suggested by the present editor.

In the concluding chapter of the volume I return to this question. However, even in the strong interpretation of the doctrine given by Professor Gormally, there *is* room for a kind of terminal sedation, where no intention to hasten death is involved. If under such circumstances 'death is hastened it is an unintended side-effect of a measure adopted purely to relieve a patient of the experience of being overwhelmed by otherwise unreleavable symptoms. That regimen is therefore clearly compatible with the doctrine of the sanctity of human life', he writes.

Chapter 9 is written by Professor Daniel Callahan at Hastings Center, a distinguished critic of euthanasia. Dr. Callahan notes that some authors have claimed that unless we give up the distinction between (actively) killing and (passively) allowing to die, we cannot consistently endorse terminal sedation. So, if we want to endorse terminal sedation, as Dr. Callahan himself wants to do, we simply have to jettison the distinction between killing and allowing to die and hence accept, together with terminal sedation, also voluntary active euthanasia. According to Dr. Callahan, we need do no such thing, however.

It is certainly true, Dr. Callahan claims, that according to existing socially constructed moral rules, we sometimes, from a moral point of view, treat cases of allowing to die as no less serious than cases of active killing. However, this is a socially constructed rule, informing us to treat some cases of allowing to die 'as if they were' cases of active killing. This does not mean that they *are* cases of active killing. To believe that they are, only because this is how we treat them, is an example of the 'artefactual fallacy', which means that we are deriving an 'is' from a (socially constructed) 'ought'. But when we realise that this is a fallacy, Dr. Callahan claims, we also realise that the distinction between killing and merely allowing to die is real.

Dr. Callahan then goes on to argue that when a terminally ill patient, who has been put into a permanent state of coma, is not artificially nourished or hydrated, what kills the patient is really the underlying disease. So this is indeed a case of allowing nature to take its course, he claims, i.e., a case of letting the patient die rather than of killing the patient. Terminal sedation, therefore, has nothing to do with euthanasia.

The subject of terminal sedation doesn't seem to go away. Ironically enough, while I was editing this volume, I happened upon a recent report about concerns that doctors in Holland may tend to resort to terminal sedation *instead* of euthanasia – in order to avoid the necessary formal procedures required in order to render euthanasia legal (*Lancet*, 19 April, 2003). This gave me the idea to ask Dr. Johannes JM van Delden to contribute a chapter on terminal sedation written from a Dutch perspective. This is Chapter 10 of the volume. In the chapter, van Delden presents recent statistics that clarifies the Dutch situation. He also defends the view that, while some forms of terminal sedation (which are in accordance with *lege artis* are not euthanasia in disguise, others are, and he rejects the idea that terminal sedation could be a compromise position in the discussion about euthanasia. Such a compromise would offer too little to those who defend voluntary euthanasia, according to van Delden.

In the concluding Chapter 11 I give some reflections on the Sanctity-of-Life Doctrine and the Active/Passive distinction. I note that, in the weak interpretation of the doctrine I suggest, it is possible for adherents of the doctrine not only to endorse terminal sedation as a kind of substitute for euthanasia, but also to endorse stopping eating and drinking as a substitute of physician assisted suicide. In my opinion, these are reasons to interpret the Sanctity-of-Life Doctrine as incorporating a reference to the active/passive distinction. These are the final words of this anthology, but certainly not the final words about this hotly contested subject.

In the appendix some authoritative views on terminal sedation have been reprinted together with the statement on euthanasia by the Sacred Congregation of Faith, a document of importance to some of the discussions in the book.

At last we find a selective bibliography with articles in English in scholarly journals since 1990 on terminal sedation. Even if the notion of 'terminal' illness is commonplace in palliative care, and even if the practise of sedation is not new within palliative care, it seems that the very notion of 'terminal sedation' is an invention of the 1990s.

3. CONCLUSION

Some questions were put to those who participated in the symposium (they were asked to comment on the theses put forward in Chapter 2). Have any answers to the queries been forthcoming? Yes, indeed. However, the answers are, not unexpectedly, conflicting.

Concerning the definition of 'terminal sedation' there has been a unanimous opinion that, in order to catch what has been controversial in the discussion, the notion should indeed include two components: (i) the terminal patient should be sedated definitely (permanently) and deeply (into coma), and (ii) when the terminally ill patient has been put into coma, artificial nutrition and hydration should be withheld. It is a moot question, however, what it means for a patient to be 'terminally ill'. Does this mean merely that the patient suffers from a disease that, in due time, will kill the patient? Or, should we also require that death is imminent?

Some find the term 'terminal sedation' a misnomer (Dr. Magna Andreen Sachs), but all participants in this volume seem to agree that words are not that important. The important thing is that we all know what we mean by them. And all who have contributed to this volume seem to agree that, on such a strict notion of terminal sedation, what has been truly controversial in the discussion is captured.

Should terminal sedation, thus (strictly) conceived, at all be an option within palliative care? No, say Dr. Magna Andreen Sachs and Professor Luke Gormally. Professor Gormally qualifies his answer, however. He can accept terminal sedation, even including the stopping of feeding the patient, but only if the decision to stop feeding the patient is not intended to hasten death.

Why should terminal sedation not be provided, according to Dr. Magna Andreen Sachs? Mainly because it is euthanasia in disguise, she submits.

Is terminal sedation euthanasia in disguise? No, the editor claims, and gets support from Professor Daniel Callahan. Yes, it is indeed euthanasia in disguise, Professor Helga Kuhse and Professor Dan Brock argue. And according to the adherent of the Sanctity-of-Life Doctrine, Professor Luke Gormally, terminal sedation and the Sanctity-of-Life Doctrine are inconsistent with each other, at least to the extent that, behind the decision not to feed the sedated patient, there is a (conditional) intention to hasten the death of the patient. The editor of the volume disagrees.

Should terminal sedation be given only as a last resort, or should it be provided at the patient's request? Only as a last resort, Dr. Gunnar Eckerdal argues. It should be provided at the patient's request, the other authors claim – with the exception of Dr. Andreen Sachs who holds that it should not be considered an option in the first place, and with the qualification from Professor Luke Gormally that the intention behind the decision to stop feeding the patient must not be to hasten death.

What about incompetent patients, should terminal sedation be provided for them? Yes, according to the editor of the volume. No, according to Dr. Magna Andreen Sachs. Yes, according to Dr. Gunnar Eckerdal. As a matter of fact, it seems as though he would claim that terminal sedation should *only* be given to incompetent patients (on the reasonable presumption that this is what they would ask for, were they in a position to ask). Patients who are competent should have a possibility to reconsider their wish to be sedated. So they should never be put *permanently* to sleep, according to Dr. Eckerdal.

The other authors do not explicitly touch upon this thorny question. This fact may be a subject for further thought. While authors who stress very much the right of the patient to control his or her life (Professors Kuhse and Brock), and who defend euthanasia at the patient's request, seem reluctant to accept that terminal sedation be provided *without* a patient's informed consent, a physician, trained and socialised in the rules of his profession, such as Dr. Gunnar Eckerdal, seems to feel, and be prepared to assume, a *special* responsibility with respect to this category of patients.

To my knowledge, this is the first book devoted exclusively to the problem of terminal sedation, a problem that has been in the focus of the discussion since the early 1990s. It is obvious that the book does not give the final verdict on the subject. It does contain strong defences by very competent thinkers of the most important and influential positions, however. I publish it in the hope that it will provoke further thought and discussion.

Torbjörn Tännsjö, December, 2003

NOTES

[1] Gillian M. Craig, 'On withholding nutrition and hydration in the terminally ill: has palliative medicine gone too far?', *Journal of medical ethics*, Vol. 20, 1994: 139-143.
[2] Cf Billings JA, Block SD, 'Slow Euthanasia', *Journal of Palliative Care* 1996;12(4):21-30.
[3] Torbjörn Tännsjö, 'Terminal Sedation: A Possible Compromise in the Euthanasia Debate?', *Bulletin of Clinical Ethics,* No. 163, November 2000, pp. 13–22

CHAPTER 1

TIMOTHY E. QUILL, BERNARD LO, AND DAN W. BROCK

PALLIATIVE OPTIONS OF LAST RESORT

A Comparison of Voluntary Stopping Eating and Drinking, Terminal Sedation, Physician-Assisted Suicide, and Voluntary Active Euthanasia

1. INTRODUCTION

PALLIATIVE CARE is the standard of care when terminally ill patients find that the burdens of continued life-prolonging treatment outweigh the benefits.[1-4] To better relieve suffering near the end of life, physicians need to improve their skills in palliative care and to routinely discuss it earlier in the course of terminal illness. In addition, access to palliative care needs to be improved, particularly for those Americans who lack health insurance. However, even the highest-quality palliative care fails or becomes unacceptable for some patients, some of whom request help hastening death. Between 10% and 50% of patients in programs devoted to palliative care still report significant pain 1 week before death.[1,5-7] Furthermore, patients request a hastened death not simply because of unrelieved pain, but because of a wide variety of unrelieved physical symptoms in combination with loss of meaning, dignity, and independence.[8,9]

How should physicians respond when competent, terminally ill patients whose suffering is not relieved by palliative care request help in hastening death? If the patient is receiving life-prolonging interventions, the physician should discontinue them, in accordance with the patient's wishes. Some patients may voluntarily stop eating and drinking (VSED). If the patient has unrelieved pain or other symptoms

Torbjörn Tännsjö (ed.), Terminal Sedation: Euthanasia in Disguise? 1-14.

and accepts sedation, the physician may legally administer terminal sedation (TS). However, it is generally legally impermissible for physicians to participate in physician-assisted suicide (PAS) or voluntary active euthanasia (VAE) in response to such patient requests. The recent Supreme Court decisions that determined that there is no constitutional right to PAS placed great emphasis on the importance of relieving pain and suffering near the end of life.[10,11] The Court acknowledged the legal acceptability of providing pain relief, even to the point of hastening death if necessary, and left open the possibility that states might choose to legalize PAS under some circumstances.

In this article, we compare VSED, TS, PAS, and VAE as potential interventions of last resort for competent, terminally ill patients who are suffering intolerably in spite of intensive efforts to palliate and who desire a hastened death. Some clinicians and patients may find some of the differences between these practices to be ethically and psychologically critical, whereas others perceive the differences as inconsequential. We will define and compare the practices, examine underlying ethical justifications, and consider appropriate categories of safeguards for whichever practices our society eventually condones.

2. DEFINITIONS AND CLINICAL COMPARISONS

With VSED, a patient who is otherwise physically capable of taking nourishment makes an active decision to discontinue all oral intake and then is gradually 'allowed to die,' primarily of dehydration or some intervening complication.[12-14] Depending on the patient's pre-existing condition, the process will usually take 1 to 3 weeks or longer if the patient continues to take some fluids. Voluntarily stopping eating and drinking has several advantages. Many patients lose their appetites and stop eating and drinking in the final stages of many illnesses. Ethically and legally, the right of competent, informed patients to refuse life-prolonging interventions, including artificial hydration and nutrition, is firmly established, and voluntary cessation of 'natural' eating and drinking could be considered an extension of that right. Because VSED requires considerable patient resolve, the voluntary nature of the action should be clear. Voluntarily stopping eating and drinking also protects patient privacy and independence, so much so that it potentially requires no participation by a physician.

The main disadvantages of VSED are that it may last for weeks and may initially increase suffering because the patient may experience thirst and hunger. Subtle coercion to proceed with the process may occur if patients are not regularly offered the opportunity to eat and drink, yet such offers may be viewed as undermining the patient's resolve. Some patients, family members, physicians, or nurses may find the notion of 'dehydrating' or 'starving' a patient to death to be morally repugnant. For patients whose current suffering is severe and unrelievable, the process would be unacceptable without sedation and analgesia. If physicians are not involved, palliation of symptoms may be inadequate, the decision to forgo eating and drinking may not be informed, and cases of treatable depression may be missed. Patients are likely to lose mental clarity toward the end of this process, which may undermine their sense of personal integrity or raise questions about whether the action remains voluntary.

Although several articles,[12,13] including a moving personal narrative,[14] have proposed VSED as an alternative to other forms of hastened death, there are no data about how frequently such decisions are made or how acceptable they are to patients, families, physicians, or nurses.

With TS, the suffering patient is sedated to unconsciousness, usually through ongoing administration of barbiturates or benzodiazepines. The patient then dies of dehydration, starvation, or some other intervening complication, as all life-sustaining interventions are withheld.[15-18] Although death is inevitable, it usually does not take place for days or even weeks, depending on clinical circumstances. Because patients are deeply sedated during this terminal period, they are believed to be free of suffering.

It can be argued that death with TS is 'foreseen' but not 'intended' and that the sedation itself is not causing death.[15-18] The sedation is intended to relieve suffering, a long-standing and uncontroversial aim of medicine, and the subsequent withholding of life-sustaining therapy has wide legal and ethical acceptance. Thus, TS probably requires no change in the law. The recent Supreme court decision gave strong support to TS, saying that pain in terminally ill patients should be treated, even to the point of rendering the patient unconscious or hastening death.[10-11] Terminal sedation is already openly practised by some palliative care and hospice groups in cases of unrelieved suffering, with a reported frequency from 0% to 44% of cases.[1,6,7,15-20]

Terminal sedation has other practical advantages. It can be carried out in patients with severe physical limitations. The time delay between initiation of TS and death permits second-guessing and reassessment by the health care team and the family. Because the health care team must administer medications and monitor effects, physicians can ensure that the patient's decision is informed and voluntary before beginning TS. In addition, many proponents believe that it is appropriate to use TS in patients who lack decision-making capacity but appear to be suffering intolerably, provided that the patient's suffering is extreme and otherwise unrelievable, and the surrogate or family agrees.

Nonetheless, TS remains controversial[21-21] and has many of the same risks associated with VAE and PAS. Like VAE, the final actors are the clinicians, not the patient. Terminal sedation could therefore be carried out without explicit discussions with alert patients who appear to be suffering intolerably or even against their wishes. Some competent, terminally ill patients reject TS. They believe that their dignity would be violated if they had to be unconscious for a prolonged time before they die, or that their families would suffer unnecessarily while waiting for them to die. Patients who wish to die in their own homes may not be able to arrange TS because it probably requires admission to a health care facility. There is some controversy in the anaesthesia literature about whether heavily sedated persons are actually free of suffering or simply unable to report or remember it.[24-26] In some clinical situations, TS cannot relieve the patient's symptoms, as when a patient is bleeding uncontrollably from an eroding lesion or a refractory coagulation disorder, cannot swallow secretions because of widespread oropharyngeal cancer, or has refractory diarrhoea from the acquired immunodeficiency syndrome (AIDS). Although such patients are probably not conscious of their condition once sedated, their death is unlikely to be dignified or remembered as peaceful by their families. Finally, and perhaps most critically, there may be confusion about the physician's ethical responsibility for contributing to the patient's death.[21,22]

With PAS, the physician provides the means, usually a prescription of a large dose of barbiturates, by which a patient can end his or her life.[1,3,27] Although the physician is morally responsible for this assistance, the patient has to carry out the final act. Physician-assisted suicide has several advantages. For some patients, access to a lethal dose of medication may give them the freedom and reassurance to continue living, knowing they can escape if and when they choose.[28,29] Because patients have to ingest the drug by their own hand, their action is likely to be voluntary. Physicians report being more comfortable with PAS than VAE,[30-32] presumably because their participation is indirect.

Opponents of PAS believe that it violates traditional moral and professional prohibitions against intentionally contributing to a patient's death. Physician-assisted suicide also has several practical disadvantages. Self-administration does not guarantee competence or voluntariness. The patient may have impaired judgment at the time of the request or the act or may be influenced by external pressures. Physician-assisted suicide is limited to patients who are physically capable of taking the medication themselves. It is not always effective,[33,34] so families may be faced with a patient who is vomiting, aspirating, or cognitively impaired, but not dying. Patients brought to the emergency department after ineffective attempts are likely to receive unwanted life-prolonging treatment. Requiring physicians to be present when patients ingest the medication could coerce an ambivalent patient to proceed, yet their absence may leave families to respond to medical complications alone.

Physician-assisted suicide is illegal in most states, but no physicians have ever been successfully prosecuted for their participation.[3] Several studies have documented a secret practice of PAS in the United States. In Washington Sate, 12% of physicians responding to a survey had received genuine requests for PAS within the year studied.[8] Twenty-four percent of requests were acceded to, and over half of those patients died as a result. An Oregon study showed similar results.[35] Physician-assisted suicide is usually conducted covertly, without consultation, guidelines, or documentation. Public controversy about legalizing PAS continues in the United States. After narrow defeats of referenda in the states of Washington and California, an Oregon referendum was passed in 1994 that legalized PAS, subject to certain safeguards.[36] After a series of legal challenges, the Oregon legislature required that the referendum be resubmitted to the electorate this November before implementation, and it was repassed this November by a margin of 60% to 40%. The US supreme Court ruled that laws in the states of Washington and New York prohibiting PAS were not unconstitutional, but the court simultaneously encouraged public discussion and state experimentation through the legislative and referendum processes.[10,11,37,38]

With VAE, the physician not only provides the means, but is the final actor by administering a lethal injection at the patient's request.[1,3,25] As practised in the Netherlands, the patient is sedated to unconsciousness and then given a lethal injection of a muscle-paralysing agent like curare. For patients who are prepared to die because their suffering is intolerable, VAE has the advantages of being quick and effective. Patients need not have manual dexterity, the ability to swallow, or an intact gastrointestinal system. Voluntary active euthanasia also requires active and direct physician participation. Physicians can ensure the patient's competence and voluntariness at the time of the act, support the family, and respond to complications. The directness of the act makes the physician's moral responsibility clear.

On the other hand, VAE explicitly and directly conflict with traditional medial prohibitions against intentionally causing death.[39] Although intended to relieve suffering, VAE achieves this goal by causing death. Furthermore, VAE could be conducted without explicit patient consent.[40,41] If abused, VAE could then be used on patients who appear to be suffering severely or posing extreme burdens to physician, family, or society, but have lost the mental capacity to make informed decisions.

The Netherlands is the only country where VAE and PAS are openly practised, regulated, and studied, although the practices remain technically illegal. According to the Remmelink reports,[9,42,43] VAE accounts for 1.8% to 2.4% of all deaths, and PAS, another 0.2% to 0.4%. In 0.7% to 0.8% of deaths, active euthanasia was performed on patients who had lost the capacity to consent, raising concern about whether guidelines restricting VAE to competent patients can be enforced in practice.[44]

United States laws prohibiting VAE, however, are stricter than those governing PAS and more likely to be prosecuted. Physicians are also more reluctant to participate in VAE even if it were legalized.[30,31] Even less is known about the secret practice of VAE than of PAS in the United States. The recent Washington State study showed that 4% of physicians had received a genuine request for VAE within the year studied, and 24% of those responded by administering a lethal injection.[8] Voluntary active euthanasia was recently legalized in a province of Australia, but this legalization was subsequently reversed by the legislature.[45]

3. ETHICAL COMPARISONS BETWEEN THE PRACTICES

Many normative ethical analyses use the doctrine of double effect and the distinction between active and passive assistance to distinguish between currently permissible acts that may hasten death (forgoing life-sustaining treatment and high-dose pain medications) and those that are impermissible (PAS and VAE).[1,2,4,46,47] Both TS and VSED have been argued to be ethically preferable alternatives to PAS and VAE on the basis of similar arguments.[12,13,16,19] In this section, we will critically examine these analyses. We also discuss the issues of voluntariness, proportionality, and conflict of duties, which may ultimately be more central to the ethical evaluation of these options. We suggest that there are more problems with the doctrine of double effect and the active/passive distinction than are ordinarily acknowledged and that TS and

VSED are more complex and less easily distinguished ethically from PAS and VAE than proponents seem to realize. Our discussion in this section will be restricted to the potential ethical permissibility of these actions and not the public policy implications.

3.1 Doctrine of Double Effect

When evaluating an action, the doctrine of double effect distinguishes between effects that a person intends (both the end sought and the means taken to the end) and consequences that are foreseen but unintended.[21,22,48,49] As long as the physician's intentions are good, it is permissible to perform actions with foreseeable consequences that it would be wrong to intend. In this view, intentionally causing death is morally impermissible, even if desired by a competent patient whose suffering could not otherwise be relieved. But if death comes unintentionally as the consequence of an otherwise well-intentioned intervention, even if foreseen with a high probability, the physician's action can be morally acceptable. The unintended but foreseen bad effect must also be proportional to the intended good effects.

The doctrine of double effect has been important in justifying the use of sufficient pain medications to relieve suffering near the end of life.[1,2,4,46,47] When high-dose opioids are used to treat pain, neither the patient nor the physician intends to accelerate death, but they accept the risk of unintentionally hastening death in order to relieve the pain. The doctrine of double effect has also been used to distinguish TS from PAS and VAE.[15,16,18,19] Relief of suffering is intended in all 3 options, but death is argued to be intended with PAS and VAE but is merely foreseen with TS. Yet to us it seems implausible to claim that death is unintended when a patient who wants to die is sedated to the point of coma, and intravenous fluids and artificial nutrition are withheld, making death certain.[21,22,50] Although the overarching intention of the sedation is to relieve the patient's suffering, the additional step of withholding fluids and nutrition is not needed to relieve pain, but is typically taken to hasten the patient's wished-for death. In contrast, when patients are similarly sedated to treat conditions like status epilepticus, therapies such as fluids and mechanical ventilation are continued with the goal of prolonging life.

According to the doctrine of double effect, intentionally taking life is always morally impermissible, whereas doing so foreseeably but unintentionally can be permissible when it produces a proportionate good. As applied to end-of-life medical decision making, the intentions of the physician are given more moral weight than the wishes and circumstances of the patient. An alternative view is that it is morally wrong to take the life of a person who wants to live, whether doing so intentionally or foreseeably. In this view, what can make TS morally permissible is that the patient gives informed consent to it, not that the physician only foresees but does not intend the patient's inevitable death.

The issue of intention is particularly complicated because the determination of what is intended by the patient or physician is often difficult to verify and because practices that are universally accepted may involve the intention to hasten death in some cases.[21,51] Death is not always intended or sought when competent patients forgo life support; sometimes patients simply do not want to continue a particular treatment, but hope nevertheless that they can live without it. But some patients find their circumstances intolerable, even with the best of care, and refuse further life support with the intent of bringing about their death. There is broad agreement that physicians must respect such refusals, even when the patient's intention is to die.[1-4,46,47,51] However, such practices are highly problematic when analyzed according to the doctrine of double effect.

3.2 The active/passive distinction

According to many normative ethical analyses, active measures that hasten death are unacceptable, whereas passive or indirect measures that achieve the same ends would be permitted.[1,2,4,46,47,52] However, how the active/passive distinction applies to these 4 practices remains controversial.[21,27] Voluntary active euthanasia is active assistance in dying, because the physician's actions directly cause the patient's death. Stopping life-sustaining therapies is typically considered passive assistance in dying, and the patient is said to die of the underlying disease no matter how proximate the physician's action and the patient's death. Physicians, however, sometimes experience stopping life-sustaining interventions as very active.[53] For example, there is nothing psychologically or physically passive about taking someone off a mechanical ventilator who is incapable of breathing on his or her own. Voluntarily stopping eating and drinking is argued to be a variant of stopping life-sustaining therapy, and the patient is said to die of the underlying disease.[12,13] However, the notion that VSED is passively 'letting nature take its course' is unpersuasive, because patients with no underlying disease would also die if they stopped eating and drinking. Death is more a result of the patient's will and resolve than an inevitable consequence of his disease. Furthermore, even if the physician's role in hastening death is generally passive or indirect, most would argue that it is desirable to have physicians involved to ensure the patient is fully informed and to actively palliate symptoms.

Both PAS and TS are challenging to evaluate according to the active/passive distinction. Physician-assisted suicide is active in that the physician provides the means whereby the patient may take his or her life and thereby contributes to a new and different cause of death than the patient's disease. However, the physician's role in PAS is passive or indirect because the patient administers the lethal medication. The psychological and temporal distance between the prescribing and the act may also make PAS seem indirect and thereby more acceptable to physicians than VAE.[30-32] These ambiguities may allow the physician to characterize his or her actions as passive or indirect.[21,50]

Terminal sedation is passive because the administration of sedation does not directly cause the patient's death and because the withholding of artificial feedings and fluids is commonly considered passively allowing the patient to die.[15,16,19] However, some physicians and nurses may consider it very active to sedate to unconsciousness someone who is seeking death and then to withhold life-prolonging interventions. Furthermore, the notion that TS is merely 'letting nature take its course' is problematic, because often the patient dies of dehydration from the withholding of fluids, not of the underlying disease.

The application and the moral importance of both the active/passive distinction and the doctrine of double effect are notoriously controversial and should not serve as the primary basis of determining the morality of these practices.

3.3 Voluntariness

We suggest that the patient's wishes and competent consent are more ethically important than whether the acts are categorized as active or passive or whether death is intended or unintended by the physician.[54-56] With competent patients, none of these acts would be morally permissible without the patient's voluntary and informed consent. Any of these actions would violate a competent patient's autonomy and would be both immoral and illegal if the patient did not understand that death was the inevitable consequence of the action or if the decision was coerced or contrary to the patient's wishes. The ethical principle of autonomy focuses on patient's rights to make important decisions about their lives, including what happens to their bodies, and may support genuine autonomous forms of these acts.[27,52]

However, because most of these acts require cooperation from physicians and, in the case of TS, the health care team, the autonomy of participating medical professionals also warrants consideration. Because TS, VSED, PAS, and VAE are not part of usual medical practice and they all result in a hastened death, clinicians should have the right to determine the nature and extent of their own participation. All physicians should respect patients' decisions to forgo life-sustaining treatment, including artificial hydration and nutrition, and provide standard palliative care, including skilful pain and symptom management. If society permits some or all of these practices (currently TS and VSED are openly tolerated), physicians who choose not to participate because of personal moral considerations should at a minimum discuss all available alternatives in the spirit of informed consent and respect for patient autonomy. Physicians are free to express their own objections to these practices as part of the informing process, to propose alternative approaches, and to transfer care to another physician if the patient continues to request actions to hasten death that they find unacceptable.

3.4 Proportionality

The principles of beneficence and nonmaleficence obligate the physician to act in the patient's best interests and to avoid causing net harm.[52] The concept of proportionality requires that the risk of causing harm must bear a direct relationship to the danger and immediacy of the patient's clinical situation and the expected benefit of the intervention.[52,57] The greater the patient's suffering, the greater risk the physician can take of potentially contributing to the patient's death, so long as the patient understands and accepts that risk. For a patient with lung cancer who is anxious and short of breath, the risk of small doses of morphine or anxiolytics is warranted. At a later time, if the patient is near death and gasping for air, more aggressive sedation is warranted, even in doses that may well cause respiratory depression. Although proportionality is an important element of the doctrine of double effect, proportionality can be applied independently of this doctrine. Sometimes a patient's suffering cannot be relieved despite optimal palliative care, and continuing to live causes torment that can end only with death.[58] Such extreme circumstances sometimes warrant extraordinary medical actions, and the forms of hastening death under consideration in this article may satisfy the requirement of proportionality. The requirement of proportionality, which all health care interventions should meet, does not support any principled ethical distinction between these 4 options.

3.5 Conflict of Duties

Unrelievable, intolerable suffering by patients at the end of life may create for physicians an explicit conflict between their ethical and professional duty to relieve suffering and their understanding of their ethical and professional duty not to use at least some means of deliberately hastening death.[57,59] Physicians who believe they should respond to such suffering by acceding to the patient's request for a hastened death may find themselves caught between their duty to the patient as a caregiver and their duty to obey the law as a citizen.[58] Solutions often can be found in the intensive application of palliative care, or within the currently legitimized options of forgoing life supports, VSED, or TS. Situations in which VSED or TS may not be adequate include terminally ill patients with uncontrolled bleeding, obstruction from nasopharyngeal cancer, and refractory AIDS diarrhea or patients who believe that spending their last days iatrogenically sedated would be meaningless, frightening, or degrading. Clearly the physician has a moral obligation not to abandon patients with refractory suffering;[60] hence, those physicians who could not provide some or all of these options because of moral or legal reservations should be required to search assiduously with the patient for mutually acceptable solutions.

4. SAFEGUARDS

In the United States, health care is undergoing radical reform driven more by market forces than by commitments to quality of care,[61,62] and 42 million persons are currently uninsured. Capitated reimbursement could provide financial incentives to encourage terminally ill patients to hasten their deaths. Physicians' participation in hastening death by any of these methods can be justified only as a last resort when standard palliative measures are ineffective or unacceptable to the patient.

Safeguards to protect vulnerable patients from the risk of error, abuse, or coercion must be constructed for any of these practices that are ultimately accepted. These risks, which have been extensively cited in the debates about PAS and VAE,[39-41] also exist for TS and VSED. Both TS and VSED could be carried out without ensuring that optimal palliative care has been provided. This risk may be particularly great if VSED is carried out without physician involvement. In TS, physicians who unreflectively believe that death is unintended or that it is not their explicit purpose may fail to acknowledge the inevitable consequences of their action or their responsibility.

The typical safeguards proposed for regulating VAE and PAS[63-66] are intended to allow physicians to respond to unrelieved suffering, while ensuring that adequate palliative measures have been attempted and that patient decisions are autonomous. These safeguards need to balance respect for patient privacy with the need to adequately oversee these interventions. Similar professional safeguards should be considered for TS and VSED, even if these practices are already sanctioned by the law. The challenge of safeguards is to be flexible enough to be responsive to individual patient dilemmas and rigorous enough to protect vulnerable persons.

Categories of safeguards include the following:

1. Palliative care ineffective: Excellent palliative care must be available yet insufficient to relieve intolerable suffering for a particular patient.

2. Informed consent: Patients must be fully informed about and capable of understanding their condition and treatment alternatives (and their risks and benefits). Requests for a hastened death must be patient initiated, free of undue influence, and enduring. Waiting periods must be flexible, depending on the nearness of inevitable death and the severity of immediate suffering.

3. Diagnostic and prognostic clarity: Patients must have a clearly diagnosed disease with known lethality. The prognosis must be understood, including the degree of uncertainty about outcomes (ie, how long the patient might live).

4. Independent second opinion: A consultant with expertise in palliative care should review the case. Specialists should also review any questions about the patient's diagnosis or prognosis. A psychiatrist should consult if there is uncertainty about treatable depression or about the patient's mental capacity.

5. Documentation and review: Explicit processes for documentation, reporting, and review should be in place to ensure accountability.

The restriction of any of these methods to the terminally ill involves a trade-off. Some patients who suffer greatly from incurable, but not terminal, illnesses and who are unresponsive to palliative measures will be denied access to a hastened death and forced to continue suffering against their will. Other patients whose request for a hastened death is denied will avoid a premature death because their suffering can subsequently be relieved with more intensive palliative care. Some methods (eg. PAS, VAE, TS) might be restricted to the terminally ill because of current inequities of access, concerns about errors and abuse, and lack of experience with the process. Others (eg, VSED) might be allowed for those who are incurably ill, but not imminently dying, if they meet all other criteria, because of the inherent waiting period, the great resolve that they require, and the opportunity for reconsideration. If any methods are extended to the incurably, but not terminally, ill, safeguards should be more stringent, including substantial waiting periods and mandatory assessment by psychiatrists and specialists, because the risk and consequences of error are increased.

We believe that clinical, ethical, and policy differences and similarities among these 4 practices need to be debated openly, both publicly and within the medical profession. Some may worry that a discussion of the similarities between VSED and TS on the one hand and PAS and VAE on the other may undermine the desired goal of optimal relief of suffering at the end of life.[40,41] Others may worry that a critical analysis of the principle of double effect or the active/passive distinction as applied to VSED and TS may undermine efforts to improve pain relief or to ensure that patient's or surrogate's decisions to forgo unwanted life-sustaining therapy are respected.[67] However, hidden, ambiguous practices, inconsistent justifications, and failure to acknowledge the risks of accepted practices may also undermine the quality of terminal care and put patients at unwarranted risk.

Allowing a hastened death only in the context of access to good palliative care puts it in its proper perspective as a small but important facet of comprehensive care for all dying patients.[1-4] Currently, TS and VSED are probably legal and are widely accepted by hospice and palliative care physicians. However, they may not be readily available because some physicians may continue to have moral objections and legal fears about these options. Physician-assisted suicide is illegal in most states, but may be difficult, if not impossible, to successfully prosecute if it is carried out at the request of an informed patient. Voluntary active euthanasia is illegal and more likely to be aggressively prosecuted if uncovered. In the United States, there is an underground, erratically available practice of PAS and even VAE that is quietly condoned.

Explicit public policies about which of these 4 practices are permissible and under what circumstances could have important benefits. Those who fear a bad death would face the end of life knowing that their physicians could respond openly if their worst fears materialize. For most, reassurance will be all that is needed, because good palliative care is generally effective. Explicit guidelines for the practices that are deemed permissible can also encourage clinicians to explore why a patient requests hastening of death, to search for palliative care alternatives, and to respond to those whose suffering is greatest. [58,60,68-70]

REFERENCES

1. Foley KM. Pain, physician-assisted suicide, and euthanasia. *Pain Forum.* 1995;4:163-178.
2. Council on Scientific Affairs, American Medical Association. Good care of the dying patient. *JAMA.* 1996;275:474-478.
3. Quill TE. *Death and Dignity: Making Choices and Taking Charge.* New York, NY: WW Norton & Co; 1993:1-255.
4. American Board of Internal Medicine End of Life Patient Care Project committee. *Caring for the Dying: Identification and Promotion of Physician Competency.* Philadelphia, Pa: American Board of Internal Medicine; 1996.
5. Kasting GA. The nonnecessity of eutahasia. IN: Humber JD, Almeder RF, Kasting GA, eds. *Physician-Assisted Death.* Totowa, NJ: Humana Press; 1993:25-43.
6. Coyle N, Adelhardt J, Foey KM, Portenoy RK. Character of terminal illness in the advanced cancer patient. *J Pain Symptom Manage.* 1990;5:83-93.
7. Ingham J, Portenoy R. Symptom assessment. *Hematol Oncol Clin North Am.* 1996;10:21-39.
8. Back AL, Wallace JI, Starks HE, Pearlman RA. Physician-assisted suicide and euthanasia in Washingtong State. *JAMA.* 1996;275:919-925.
9. vanderMaas PJ, vand Delden JJM, Pijnenborg L. *Euthanasia and Other Medical Decisions Concerning the End of Life.* Amsterdam, the Netherlands: Elsevier; 1992.
10. *Vacco v Quill,* 117 SCt 2293 (1997).
11. *Washington vs Glucksberg,* 117SCt 2258 (1997).
12. Bernat JL, Gert B, Mogielnicki RP. Patient refusal of hydration and nutrition. *Arch Intern Med.* 1993;153:2723-2727.
13. Printz LA. Terminal dehydration, a compassionate treatment. *Arch Intern med.* 1992;152:697-700.
14. Eddy DM. A converesation with my mother. *JAMA.* 1994;272:179-181.
15. Chernely NI, Portenoy RK. Sedation in the management of refractory symptoms: guidelines for evaluation and treatment. *J Palliat Care.* 1994;10:31-38.
16. Troug RD, Berde DB, Mitchell C, Grier HE. Barbiturates in the care of the terminally ill. *N Engl J Med.* 1991;327:1678-1681.
17. Enck RE. *The Medical Care of Terminally Ill patients.* Baltimore, Md: Johns Hopkins University Press; 1994.
18. Saunders C, Sykes N. *The Management of Terminal Malignant Disease.* 3rd ed. London, England: Hodder Headline Group; 1993:1-305.
19. Byock IR. Consciously walking the fine line: thoughts on a hospice response to assisted suicide and euthanasia. *J Palliative Care.* 1990;6:7-11.
20. Ventaffridda B, Ripamonti C, DeConno F, et al. Symptom prevalence and control during cancer patients' last days of life. *J Palliat Care.* 1990;6:7-11.
21. Brody H. Causing, intending, and assisting death. *J Clin Ethics.* 1993;4:112-117.
22. Billings JA. Slow euthanasia. *J Palliat Care.* 1996;12:21-30.
23. Orentlicher D. The Supreme Court and physician-assisted suicide: rejecting assisted suicide but embracing euthanasia. *N Engl J Med.* 1997;337:1236-1239.
24. Moerman N, Bonke B. Oosting J. Awareness and recall during general anasthesia: facts and feelings. *Anesthesiology.* 1993;79:454-464.
25. Utting JE. Awareness: clinical aspects; consciousness, awareness, and pain. In: Rosen M, Linn JN. *General Anesthesia.* London: England: Butterworts;1987:171-179.
26. Evans JM. Patient's experience of awareness during general anesthesia; consciousness, awareness and pain. In: Rosen M, Linn JN. *General Anesthesia.* London: England: Butterworts;1987:184-192.
27. Brock DW. Voluntary active euthanasia. *Hastings Cent Rep.* 1992;22:10-22.
28. Quill TE. Death and dignity. *N Engl J Med.* 1991;324:691-694.
29. Rollin B. *Last Wish.* New York, NY: Warner Books;1985.
30. Cohen JS, Fihn SD, Boyko EJ, et al. Attitudes toward assisted suicide and euthanasia among physicians in Washington State. *N Engl J Med.* 1994;331:89-94.
31. Bachman JG, Alchser KH, Koukas DJ, et al. Attitudes of Michigan physicians and the public toward legalizing physician-assisted suicide and voluntary euthanasia. *N Engl J Med.* 1996;334:303-309.
32. Duberstein PR, Conwell Y, Cox C, et al. Attitudes toward self-determined death. *J Am Geriatr Soc.* 1995;43:395-400.

33. Preston TA, Mero R. Observations concerning terminally-ill patients who choose suicide. *J Pharm Care Pain Symptom Control.* 1996;1:183-192.
34. Admiraal PV. Toepassing van euthanatica (the use of euthanatics). *Ned Tijdschr Geneeskd.* 1995;139:265-268.
35. Lee MA, Nelson HD, Tilden VP, Ganzini L, Schmidt TA, Tolle SW. Legalizing assisted suicide: views of physicians in Oregon. *N Engl J Med.* 1996;334:310-315.
36. Alpers A, Lo B. Physician-assisted suicide in Oregon: a bold experiment. *JAMA.1995;274:483-487.*
37. *Compassion in Dying v Washington,* No. 94-35534, 1966 WL 94848 (9th Cir, Mar 6, 19969.
38. *Quill vs Vacco,* No. 95-7028 (2d Cir, April 9, 1996).
39. Gaylin W, Kass LR, Pellegrino ED, Siegler M. Doictors must not kill. *JAMA.* 1988;259:2139-2140.
40. Teno J, Lynn J. Voluntary active euthanasia: the individual case and public policy. *J Am Geriatr Soc.*1991;39:827-830.
41. Kamisar Y. Against assisted suicide — even a very limited form. *Univ Detroit Mercy Law Rev.* 1995;72:735-769.
42. vanderMaas PJ, vanderWal G, Haverkate I, et al. Euthanasia, physician-assisted suicide and other medical practices involving the end of life in the Netherlands, 1990-1995. *N Eng J Med.* 1996;335:1699-1705.
43. vanderWal G, vanderMaas PJ, Bosma JM, et al. Evaluation of the notififcation procedure for physician-assisted death in the Netherlands. *N Eng J Med.* 1996;335:1706-1711.
44. Hendin H. Seduced by death. *Issues Law Med.* 1994;10:123-168.
45. Ryan CJ, Kaye M. Euthanasia in Australia: the Nothern Territory rights of the terminally ill act. *New Eng J Med* 1996;334:326-328.
46. President's Commission for the Study of Ethical Problems in Medicine and Biomedical and Behavioural Research. *Deciding to Forego Life-Sustaining Treatment: Ethical, Medical and Legal Issues in Treatment Decisions.* Washington, DC: US Government Printing Office; 1982.
47. The Hastings Center Report. *Guidelines on the Thermination of Life-Sustaning Treatment: Ethical, Medical and Legal Issues in Treatment Decisions.* Briarcliff Manor, NY: Hastings Center, 1987.
48. Marquis DBV: Four versions of the double effect. *J Med Phils.* 1991;16:515-544.
49. Kamm F. The doctrine of double effect. *J Med Philos.* 1991;16:571-585.
50. Quill TE. The ambiguity of clinical intentions. *N Eng J Med.* 1993;329:1039-1040.
51. Alpers A, Lo B. Does it make clinical sense to equate terminally ill patients who require life-sustaining interventions with those who do not? *JAMA.* 1997;277:1705-1708.
52. Beauchamp TL, Childress JF. *Principles of Biomedical Ethics.* 3rd ed. New York, NY: Oxford University Press; 1994.
53. Edwards MJ, Tolle SW. Disconnecting a ventilator at the request of a patient who knows he will die. *Ann Intern Med.* 1992;117:254-256.
54. Orentlicher D. The legalization of physician-assisted suicide. *N Eng J Med* 1996;335:663-667.
55. Drickamer MA, Lee MA, Ganzini L. Practical issues in physician-assisted suicide. *Ann Intern Med.* 1997;126:146-151.
56. Angell M. The Suypreme Court and physician-assisted suicide: the ultimate right. *N Eng J Med* 1997;336:50-53.
57. de Wachter MAM. Active euthanasia in the Netherlands. *JAMA* 1989;262:3316-3319.
58. Quill TE, Brody RV. 'You promised me I wouldn't die like this': a bad death as a medical emergency. *Arch Intern Med.* 1995;155:1250-1254.
59. Welie JVM. The medical exception: physicians, euthanasia and the Dutch criminal law. *J Med Phils.* 1992;17:419-437.
60. Quill TE, Cassel CK. Nonabandonment: a central obligation for physicians. *Ann Intern Med.* 1995;122:368-374.
61. Emanuel EJ, Brett AS. Managed competition and the patient-physician relationship. *N Eng J Med.* 1993;329:879-882.
62. Morrison RS, Meier DE. Managed care at the end of life. *Trends Health Care Law Ethics.* 1995;10:91-96.
63. Quill TE, Cassel CK. Meier DE. Care of the hopelessly ill: proposed criteria for physician-assisted suicide. *N Eng J Med.* 1992;327:1380-1384.
64. Brody H. Assisted death. *N Eng J Med.* 1992;327:1384-1388.
65. Miller FG, Quill TE, Brody H, et al. Regulating physician-assisted death. *N Eng J Med.* 1994;331:119-123.

66. Baron CH, Bergstresser C, Brock DW, et al. Statue: a model state act to authorize and regulate physician-assisted suicide. *Harvard J Legislation.* 1996;33:1-34.
67. Mount B, Morphine drips, terminal sedation, and slow euthanasia: definitions and facts, not anecdotes. *J Palliat Care.* 1996;12:31-37.
68. Lee MA, Tolle SW. Oregon's assisted-suicide vote: the silver lining. *Ann Intern Med.* 1996;124:267-269.
69. Block SD, Billings A. Patient requests to hasten death: evaluation and management in terminal care. *Arch Intern Med.* 1994;154:2039-2047.
70. Quill TE. Doctor, I want to die: will you help me? *JAMA. 1993;270:870-873.*

CHAPTER 2

TORBJÖRN TÄNNSJÖ

TERMINAL SEDATION: A SUBSTITUTE FOR EUTHANASIA?

1. OBJECTIVE

By 'terminal sedation' I denote, in the present chapter, a procedure where through heavy sedation a terminally ill patient is put into a state of coma, where the intention of the doctor is that the patient should stay comatose until he or she is dead. No extraordinary monitoring of the medical state of the patient is undertaken. Normal hydration is ignored. All this means that in some cases where patients are being terminally sedated, death is hastened; if the disease does not kill the patient, some complication in relation to the sedation, or the withdrawal of treatment and hydration, or the combination of these, does.

The moral status of terminal sedation is controversial, and in most countries the legal status of terminal sedation is far from clear; this is true in particular of Sweden. All this means that it is of importance to scrutinise examples of this practice. And the case focused on in this chapter is of particular interest, since it has been carefully documented and extensively discussed.

2. THE SWEDISH CASE

A physician, working at *Kålltorp Hospice*, Dr. Mats Holmberg, chose in two cases in 1996, and in one case in 1997, to sedate three terminally ill patients. They all died in their sleep within two to four days after a Dormicum (Midazolam) drip had been administered to them. The first patient, Mr. Winter, in his sixtieth year, suffered from a brain tumour. In his terminal state he was no longer communicable. However, he was a tall and unusually strong person and, out of confusion and anxiety, he acted aggressively. It was decided that he could no longer be treated at the hospice. He should be transferred to a closed psychiatric ward. His close ones

Torbjörn Tännsjö (ed.), Terminal Sedation: Euthanasia in Disguise?, 15-30.

protested. They felt that what Mr. Winter really wanted was to die. The doctor suggested terminal sedation, which was accepted by the close ones. The doctor claimed afterwards that he had acted on the presumption that this was really what Mr. Winter himself would have requested, had he been communicable.

The second patient, Mrs. Spring, was born in 1932. She suffered from cancer in her bladder in a terminal state. Her physical suffering had been alleviated, but she repeated a request for euthanasia she had made when she was admitted to the hospice; she found that her anxiety was unbearable and saw no further point in her life. She was informed that euthanasia was illegal, but that she might obtain intermittent sedation with Dormicum. She then demanded to be *terminally* sedated. Her husband supported her in her request, which was honoured by the physician. After having parted from her husband and children, Mrs. Spring was put to sleep and five days later she was dead.

The third patient, Mr. Autumn, was also suffering from a brain tumour. He was dysphatic but communicable. In his terminal state he developed anxiety, refused to take anti-depressant medicines, and eventually he asked for euthanasia. He was informed that euthanasia was out of the question, and he was offered intermittent sedation. Even Mr. Autumn rejected this offer. However, when his son talked through his situation with him and presented to him the possibility that he could be terminally sedated, he gratefully accepted this. The son felt that his father was competent to make the decision and he insisted that the wish of the father should be honoured. A decision was reached that the patient should be terminally sedated. However, when Dormicum was given, Dr. Holmberg was not himself present and a nurse, who actually administered the drip, had serious concerns about the treatment, and so had a colleague of hers whom she consulted. The nurse who administered the drip contacted the Swedish Association of Health Officers (her trade union), which brought the case to The National Board of Health and Welfare.

An investigation was performed by The National Board of Health and Welfare and the case was handed over to the district attorney. The district attorney dropped the case, however, on the ground that it was impossible to find out what had caused death in the three cases (no autopsy had been performed).

Before going into the arguments about the actual Swedish case, I will put what happened into perspective.

3. THE SCHOLARLY BACKGROUND

A discussion has been going on during the 1990s about the moral status of terminal sedation. Three main positions surface when the literature is examined.

The first position is that terminal sedation is never morally acceptable. This position is exemplified in an early discussion that took place in *The Journal of Clinical Ethics*, focusing on a case where a patient at his own request received light sedation before he was removed from his respirator. He received 30 mg of intravenous morphine before the ventilator was disconnected and he fell asleep, but once the ventilator had been disconnected, ‘he awoke, took several deep gasps for air, became apneic, and died several minutes later’. And still it would have been wrong to sedate him terminally, some authors taking up the first position have argued.[1]

Here is a general statement of the same position (not focusing especially on patients who are being taken off their ventilators):

> Having decided that sedation is needed, the doctor must try to find a drug regime that relieves distress but does not prevent the patient from verbal communication with friends and relatives, and does not lead to toxic side-effects, or expedite death.[2]

According to a second position, terminal sedation may be resorted to when all other attempts to assuage refractory pain, anxiety, and distress have failed, provided it is impossible in the circumstances to strike a balance between light and heavy sedation. Here is an example of *this* position, put forward by reputed hospice representatives:

> Terminal agitation must be treated aggressively, otherwise the distress of the patient will become extreme. Even when incremental doses of sedatives are given, it is rarely possible to achieve a balance between relief of agitation and alertness.[3]

And finally, according to the third position in the discussion, terminal sedation is seen as one option among others, that should be delivered on request to patients in a terminal state, suffering from refractory physical or psychological pain (and who feel themselves that sedation is preferable to opioids and light sedation). Here is an example of this position, taking the case described above as a point of departure:

> My own colleagues would probably have employed sufficient benzodiazepines to alleviate anxiety; then they would have administered repeated boluses of morphine to ensure complete sedation throughout the entire process of ventilatory withdrawal even if it required several hundred milligrams ... In other words, I think the question should be turned around: Is it morally justifiable *not* to sedate an alert patient before ventilator removal? If I were the patient described in this report I would have wanted my physicians … to have been more bold.[4]

4. STANDARD MORAL AND LEGAL THINKING

Terminal sedation is not the only possible option in the care of terminally ill patients who suffer from severe refractory pain or discomfort. In the Netherlands and in Belgium, the active and intentional killing of patients with incurable diseases, suffering in what they themselves consider to be an intolerable manner, is legally tolerated, if it takes place at the request of these patients and if a certain procedure is being followed (euthanasia). But in other countries euthanasia is legally prohibited. The most common argument against euthanasia has been that the practice of euthanasia must come to violate two basic principles of medical ethics: the principle of acts and omissions and the principle of double effect. It is therefore of interest to examine these principles and to try to find out whether they can be used to draw a distinction between, on the one hand euthanasia, which is clearly incompatible with the principles, and on the other hand terminal sedation, which is compatible with them. This would prepare the way for people adhering to these principles, and who reject euthanasia, to endorse terminal sedation in palliative care.

Take first the distinction between acts and omissions. On this idea, while it is always wrong actively to kill a person, it may sometimes be right to allow death to come about. Active killing is always wrong, passive killing may sometimes be right.

Contrary to what many medical ethicists seem to believe, the distinction is comprehensible. It is certainly true that the distinction does not allow us to say, in relation to any concrete and particular action, whether it falls into either the active or the passive category.[5] All concrete actions are active under some description of them. However, some *kinds* of actions allow that we sort instances of them into the active or passive category, relative to the kind in question. To *help* a person is an example of this. Instances of this kind (helpings) are either 'active' or 'passive', not in any absolute sense, but with respect to this kind. We can actively help people, but we can also help people passively, by just allowing that benefits befall them. And killing, or hastening death, is indeed another example of this. There are clear cut cases of active killing, and there are clear cut cases of passive killing (of allowing nature to take its course). No *criterion* can be formulated here, I think, but no criterion is really needed. Our linguistic intuitions are clear enough. In particular cases we can say of an act of killing, not if it is 'active' or 'passive' in any absolute sense, but, *qua* 'killing' whether it is active or passive, and we can even state our reasons for this assessment (although these reasons cannot take a quite general form). Not to feed a patient, who, as a consequence, starves to death, is to kill passively (to allow nature to take its course), while injecting an opioid, which kills the patient, is to kill actively.

Note that my act of actively killing a patient may consist of some actions, which we tend to describe as 'omissions' (although not with respect to the killing in question). Suppose I give a drip to a patient. I happen to give it too fast. I notice this, but I omit doing anything about it. Consequently the patient dies. This is a clear case of active killing, even though it consists also of an omission (the omission of slowing down the infusion).

Now, is the distinction between active and passive killing of moral importance? Here it must suffice to note that, in medical practice in modern Western countries the distinction between actively and passively killing as such is of no direct importance. For in most Western countries, even active killing of severely ill patients is legally tolerated. I think here of cases where patients are given sedative medication or opioids in a manner that hastens death. *This* is clearly a case of *active* killing.

What about the principle of double effect, then? According to this principle it is always wrong intentionally to kill a patient, but it may be right to provide aggressive palliative care, with the intention of relieving pain, even if it can be foreseen that the patient will die from the care in question. This may be right provided that, in the circumstances, it is a good thing to have the patient free of pain, and, provided there is some reasonable proportion between the (first) good effect (the patient being free of pain) and the (second) bad effect (death being somewhat hastened).[6]

It is certainly true that it may be very difficult to *tell* whether a certain doctor in a certain situation is intending the death of a patient, when giving the patient opioids in a manner that hastens death (this may be difficult to tell even for the doctor him- or herself), but, *in principle*, there is a clear difference between a case where the doctor gives the medication, intending to kill the patient, because this is considered the best way to relieve the patient's pain and, on the other hand, a case where the doctor administers the medication in order to relieve pain (realising that the patient may get killed, as an unwanted and unavoidable consequence of this).

A way of checking whether, according to this principle, the intention of the doctor is the right one is to ask the doctor: if you could have relieved the pain in another way, that had not hastened death, would you have done so? If the answer to this question is in the negative, then the doctor has not abided by the principle of double effect. If the answer is positive, then the doctor has abided by it.

Now, even if the principle of double effect is reasonably clear and comprehensible, it should be noted that in contemporary Western countries, we do not abide by it. In most Western countries intentional killing of patients is legally tolerated. The much publicised Bland case bears witness to this.[7] In this case it was decided that a patient in a persistent vegetative state should not be artificially fed or hydrated any more. It is *obvious* that the intention behind the action (of not feeding or hydrating the patient) was to hasten death. Similar cases are easily found in most other Western countries, including Sweden. And they are all legally tolerated.

Does all this mean that neither the acts and omissions principle nor the principle of double effect plays any role in Western countries, and that the avenue to euthanasia is open? No, this is clearly not the case. However, only a *combination* of the two principles (of acts and omissions and the double effect) can substantiate the traditional approach. Note that this is how the legal situation in most Westerns countries can be represented (putting the Netherlands and Belgium to one side):

KILLING	Death intended	Death merely foreseen
Active	FORBIDDEN	TOLERATED
Passive	TOLERATED	TOLERATED

Figure

This may seem strange. If active killing is allowed, and if intentional killing is allowed, what is so problematic about their combination? Must not a simple logical fallacy lie behind the rejection of euthanasia?

No, on the contrary, it is the argument from the observation that a certain distinction lacks moral relevance in one situation to the conclusion that it lacks relevance in all situations that is fallacious. This has been observed in particular by Shelly Kagan.[8] The standard ethical thinking on this point, no matter what we may think about it, is consistent. And the rationale behind it could be stated roughly as follows: while it is not morally problematic, as such, to kill passively (to allow that nature takes its course), it *is* morally problematic as such to kill actively. Of course, this does not mean that *all* instances of passive killing are morally acceptable. Sometimes it is morally wrong to kill passively. As a matter of fact, this is wrong, and very wrong, in most cases. But when it is wrong to kill passively, this is not due to any *inherent* wrongness in the act, but to particular *consequences* of it. It may for example be wrong to allow a patient to die because of lack of treatment, if one has promised, or undertaken, to provide the treatment in question, most obviously so if the treatment would had saved the patient.

But to condemn all kinds of active killing, irrespective of the consequences, may be to go too far. Here a further qualification may be needed, and we can make it with reference to the principle of double effect. What is *inherently* wrong is only active killing *with the intention to kill* (actively). If the killing is active, but death merely a foreseen consequence of the act in question, then the action may be right. However, actively and deliberately to kill a patient is wrong, period, i.e., it is wrong irrespective of the consequences.

This is standard thinking, then, and it is also a kind of thinking which can be endorsed by what has been called the Sanctity-of-Life Doctrine.

Now, if this characterisation of standard thinking (and the implications of the Sanctity-of-Life Doctrine) is correct, where does this leave terminal sedation? If, in accordance with standard thinking and the Sanctity-of-Life Doctrine, we hold on to the absolute prohibition against active and intentional killing of patients, does this mean that, together with voluntary active and intended euthanasia, we must reject terminal sedation as well?

5. TERMINAL SEDATION, MORALITY AND THE LAW

As far as I can see, terminal sedation is compatible with standard thinking (and the Sanctity-of Life Doctrine) within medical ethics. To see this we should contemplate the following line of argument. When a patient is being terminally sedated, and having all treatment withdrawn, this act falls outside the 'forbidden' top left box in the figure above. The sedation can hasten death, of course. But, the point in sedating the patient is not to cause death, but to relieve suffering. So even if the sedation (actively) kills the patient, the death of the patient is merely foreseen, not intended. On the other hand, the withdrawal of treatment is undertaken with the intention to hasten the death of the patient. However, this is a case of passive, not of active, killing. Death is certainly sought, but not actively, only as the end of a natural chain of events. Death is allowed to take place in the way it does. A similar line of argument has been adopted by the American Association for Hospice and Palliative Medicine, in a position statement on the end-of-life, where the decision to withhold nutrition is clearly separated from the decision to sedate the patient:

> Patients for whom sedation may be appropriate are most often near death as a result of an underlying disease process. Although the withdrawal of artificial hydration and nutrition commonly accompanies sedation, the decision to provide, withdraw, or withhold such treatments is separate from the decision whether or not to provide sedation.[9]

This means that terminal sedation, in contradistinction to active and intended euthanasia, avoids the forbidden top left box in the above figure.

However, this line of argument has been questioned. Timothy Quill, Bernard Lo, and Dan Brock argue in an article reprinted as Chapter 1 of this book that terminal sedation is much more like active and voluntary euthanasia than it might at first appear. This is how these authors argue:

> Yet, to us it seems implausible to claim that death is unintended when a patient who wants to die is sedated to the point of coma, and intravenous fluids and artificial nutrition are withheld, making death certain.[10]

However, the authors do not distinguish between on the one hand sedation (where death is produced by active means, but where death is not intended), and on the other hand the withholding of hydration and nutrition (where death is intentionally sought, but sought by passive means).

But is it true that withholding life-sustaining treatment is passive (and not active killing)? The authors even doubt this (with reference to how they believe that physicians see the case):

> Physicians, however, sometimes experience stopping life-sustaining interventions as very active. For example, there is nothing psychologically or physically passive about taking someone off a mechanical ventilator who is incapable of breathing on his or her own.[11]

But while it is true that taking a patient off a mechanical ventilator is a very active action indeed (of *taking someone from a mechanical device*), this does not mean that it is a case of active *killing*. Remember that all concrete actions are active under, and relative to, some description of them, but remember also that, in many cases, they are passive under, and relative to, some other description. The taking someone off a mechanical ventilator is a clear example of this. It is a case of passive killing.

Here we may indeed have recourse to the notion that, once the patient has been taken off the ventilator, nature 'takes its course'. To this the authors raise the following objection, however:

> [T]he notion that TS [terminal sedation] is merely 'letting nature take its course' is problematic, because often the patient dies of dehydration from the withholding of fluids, not of the underlying disease.[12]

However, it is no less natural to die from thirst or hunger (if one gets neither fluids nor nourishment) than to die from a disease. Nature takes its course in *both* cases.

But ought not the physician to provide life-sustaining treatment, once the patient has been terminally sedated? I have come across one argument to this effect. Dr. Gillian Craig has argued in the following manner:

> Where the side effects of a given treatment such as sedation are predictable, lethal, and easily overcome by simple measures such as intravenous fluids to prevent dehydration, failure to use such measures could be regarded as negligence and warrants a conviction for manslaughter. These days the use of intravenous fluids cannot be regarded as an extraordinary measure. It is not however a measure that can be continued for more than a week or so, but if the patient is truly terminally ill, death from natural causes will intervene before problems with drip maintenance arise. There are other routes by which fluids can be given if need be.[13]

Is there a point in abiding by this practice? No, at least not if a quick and smooth death is what is requested by the patient him- or herself. A patient has both a legal and moral right to starve to death. Then a patient must also have a right, in order to hasten death, to refuse intravenous fluids.

But what if the patient, like Mr. Winter in the Swedish case, is not competent? Then, I submit, the patient should be treated on the presumption that the treatment he or she receives is the treatment he or she *would* have requested, had he or she been competent. So if the doctor who sedated Mr. Winter is correct when he claims that this is what both he and Mr. Winter's close ones thought was in Mr. Winter's best interest, and what Mr. Winter himself would have requested, had he been competent, then the doctor did the right thing when he terminally sedated Mr. Winter.

However, what the patient would have requested may be hard to find out. And it is certainly true that we can never know this for certain. Is it therefore not reasonable to think that, in the circumstances, there exists at least a presumption in favour of hydrating the patient? This is what Craig argues:

> … although doctors should not give treatment simply because it is available, in cases of doubt about the best interests of the patient the presumption should be in favour of prolonging life.[14]

Now, it is certainly true that, if the prognosis is considered uncertain, then there is a point in presuming that prolonged life is what the patient would have requested. However, if the patient is dying, there is no point in this presumption. Certainly, to hasten death, when this goes against the will of the patient, means manslaughter. However, to prolong life, when *this* goes against the will of the patient, means torture. I cannot see that torture is any better than manslaughter. So there is no way of 'erring on the right side' here.

6. EXPERT REPORTS

When The National Board of Health and Welfare filed its complaint, it based its accusation on reports written by medical and legal experts. All the medical experts take up a hard line and argue that terminal sedation should never take place. Dormicum may be used in some cases, but the intention should always be to keep the patient communicable. How do they substantiate this point?

Professor Peter Strang, who holds the only Swedish chair in Palliative Care Medicine, argues that terminal sedation means that the distinction between standard palliative care and (illegal) euthanasia had been transgressed. His argument was repeated by Associated Professor Magna Andreen Sachs, who is a member of the scientific board of The National Board of Health and Welfare, and by Professor Barbro Beck-Friis, a retired doctor, who has been a pioneer and leading in the field of palliative medicine in Sweden. Beck-Friis even ventured the legal opinion that what had happened to Mr. Winter was 'comparable to murder'.[15]

One of the legal experts, however, the judge Staffan Vängby, seems to accept terminal sedation at the request from competent patients. He does raise doubts in relation to one of the patients, i.e., Mr. Winter, however, who had been sedated without himself having made any request to this effect. How do we know that this was in accordance with what he would have wanted, Vängby asks. Now, it is certainly true that we do not know for sure what Mr. Winter would have wanted. However, from what has been reported about the case, it is reasonable to conclude, *I* think, that the physician made the right presumption and, therefore, the right decision, even in this case. I have already given my reason for this conclusion.

The other legal expert, Suzanne Wennberg, Professor of Criminal Law, repudiated the doctrine of double effect. She concedes that in certain authoritative Swedish documents, this doctrine is taken for granted. However, as a matter of fact, the crucial thing is not the intention of the doctor but if the treatment given is part of an accepted medical practice. And if it is true that terminal sedation is not part of such an accepted practice, as has been claimed by all the medical experts, then it is illegal, she concludes.

Now, her repudiation of the doctrine of double effect is very speculative. However, let us grant her the assumption that what is decisive is established medical practice. How can it be decided what the medical practice is? This depends both on how physicians act, and on how the experts judge their acts. Where does this leave the case under scrutiny? Did the physician, who put his terminally ill patients to sleep until they were dead, conform to an accepted practice?

Well, the judgements by the medical experts called upon to give testimony in this case indicate that his actions were at variance with medical practice. However, in a discussion note in a Swedish newspaper two other experts, specialised in pain control, claimed that only at their clinic, as a last resort where symptom control failed, some ten patients were terminally sedated each year.[16] So one may perhaps claim that, even on this speculative interpretation of the legal situation, the case is unsettled.

We have seen that the expert reports add little to the traditional discussion on terminal sedation. However, there is one important remark made by Peter Strang in his report. If the dying patient is in a process of anxiety and existential pain, then, according to Strang, 'this is a process that the patient and his close ones, aided by the personnel, has to endure'.

What does 'has to' mean in this context? It refers to some kind of necessity. But it is hardly a physical necessity, since there exists a way of sparing the patient this experience: through terminal sedation.

The necessity must then be a *moral* one. However, if it is, then we have no *argument* here for the conclusion that terminal sedation should not be provided, but rather the *conclusion* of such an argument. So where are the premises? They seem to be lacking.

When it was clear that the district attorney had dropped the case, a discussion took place in *Läkartidningen* (published weekly by the Swedish Medical Association). One original argument to the effect that terminal sedation should never take place was forthcoming in this discussion. This argument was put forward by Associate Professor Magna Andreen Sachs, who had already touched upon it in her report to The National Board of Health and Welfare. According to Andreen Sachs, since no monitoring of the patients in the case here under scrutiny had taken place and no hydration had been given, the doctor's intention *must* have been to hasten death. Otherwise, life-sustaining measures would have been taken.

Provided that, if the patient is not hydrated and carefully monitored, then the risk that the sedation will kill the patient increases (if this is so is a medical question I feel incompetent to answer), then this argument may seem to be on the right track. Remember that one way of finding out whether the doctor has the appropriate intention is to ask whether the doctor would have opted for less risky measures, had such measures existed. However, when putting forward her argument in the present context the author seems to forget that, if these other (less risky) measures are at variance with the will of the patient (as expressed by the patient or as presumed by the doctor), they *cannot* be taken by a doctor who wants to respect the autonomy of his or her patient. Thus, the doctor's not taking such measures in the circumstances (where they are not allowed by the patient) does not tell us anything at all about the intention behind the doctor's decision to sedate the patient into oblivion.

7. TERMINAL SEDATION AT THE PATIENT'S REQUEST?

We have seen that both the experts that were consulted and the doctors who took part in the public discussion in the aftermath of the controversial Swedish cases of terminal sedation, tended to hold that terminal sedation should *never* be an option in palliative care. If possible, and irrespective of the will of the patient, the patient should be kept communicable. However, no good arguments were forthcoming in support of this position. And there were several Swedish doctors who went against the current and supported the use of terminal sedation. It is reasonable to assume that they adhere to what must be considered Anglo-Saxon standard hospice ideology: at least as a last resort, a terminally ill patient may be sedated into oblivion. But why only as a last resort? I now turn to this last question.

We can distinguish between three lines of argument here. First of all, it has been argued that sedation into oblivion is special; it cannot be considered as merely one reasonable alternative in palliative care among others. To request sedation into oblivion may in some cases be like asking for life threatening surgery for a disease that could easily be cured with a drug with no known side effects. No doctor can be required to provide such surgery.

Secondly, it has been argued that while terminal sedation may be a standard palliative option when death is *unavoidably* imminent, it is not a proper option when death is *rendered* imminent only through some choice made by a terminally and critically ill patient (he or she wants to be taken off a ventilator, say). A patient should never, by being offered sedation before the treatment is withdrawn, be lured into a decision to forgo life-sustaining treatment.

Finally, it has been argued that, while it is perfectly in order to sedate a patient who suffers from refractory *physical* pain and does not want to experience the rest of her life, it is less obvious that a patient who suffers from *psychological* pain should have the same right. I discuss these three kinds of argument in order.

8. IS TERMINAL SEDATION SPECIAL?

The idea that sedation into oblivion is special must rest on some idea to the effect that this kind of treatment is, in the circumstances, and from the point of view of the physician, very complicated, or, from the point of view of the patient, much too risky.

Now, I do not deny that it may be complicated terminally to sedate a patient. The physician, who is typically not a trained anaesthesiologist, must strike a balance and not give too much of the sedative (or give it too fast), or give too little (or too slowly). The first mistake may mean that death is hastened. The latter mistake may mean that the patient has to wake up, only to be sedated again. In the Swedish case no serious complications seem to have occurred. However, in the much discussed Australian case, where Dr. Philip Nitschke, who was no longer able to end his patient's life by euthanasia, provided terminal sedation, the latter kind of complication did occur. This is how the case has been reported:

> Over three days, massive doses of morphine, midazolam, ketamine, chloropromozine and barbiturates were used with the intention of keeping this woman unconscious. Despite such efforts, she woke three times, and immediately further sedatives were administered. Management was documented on national television. Death was from broncho-pneumonia, carcinomatosis and mixed drug overdose.[17]

There is obviously no way for a doctor to *guarantee* that a patient will stay comatose until death arrives. The risk to wake up, only to be sedated once again, is something a patient, who wants to undergo terminal sedation, must be prepared to accept. The way for the doctor to minimise the risk, however should be to try to err on the right side. But this means an increased risk that death is hastened through sedation. Is that acceptable?

Well, how dangerous a treatment could reasonably be offered depends on the available alternatives. We must here assess both possible gains and losses. If death is imminent, if the patient feels that there are no further objectives to pursue in life, and wants to die sooner rather than later, then I think it fair to say that terminal sedation is innocuous. It can in no way harm the patient.

As a matter of fact, *terminal* sedation here compares favourably with *other* kinds of optional anaesthesia. Even in cases where general anaesthesia, as compared to local anaesthesia, means a threat to the health and life of the patient, the patient is allowed to make an informed choice. This is how a standard Swedish textbook in anaesthesia describes the situation.

> The patient's right to choose is an important principle in anaesthesiology. In accordance with the spirit of the Law of Health Care and Medicine, the patient shall, after having been carefully informed, and to the extent this is possible, be granted an opportunity to choose between, for example, general anaesthesia (to be asleep), and local anaesthesia (to be awake). In order to render possible for the patient to exercise this right, the surgeon and the anaesthesiologist must provide, in particular with respect to medical risks, comprehensive and objective information about possible advantages and drawbacks of the options.[18]

In these cases patients are being allowed to take a considerable risk to lose *very much* (the rest of their lives), when they choose general anaesthesia instead of local anaesthesia. There is no parallel to this in terminal sedation. It is certainly true that patients who choose to become sedated into oblivion may lose the rest of their lives, but very little is left of their lives anyway, and the rest of their lives is, by their own lights, of *negative* value to them!

9. WHAT IF DEATH IS NOT IMMINENT?

However, what if the patient is critically ill but death not imminent? What about a patient, who, connected to a ventilator can live for several weeks, but who wants to be (first) terminally sedated and (then) disconnected from the ventilator? According to the second line of argument, it would be wrong to offer terminal sedation to such a patient, at least *before* the ventilator is disconnected. While the patient has a right to be taken off the ventilator it would be wrong to render such a choice more palpable to the patient, by also offering sedation. This position was taken by Robert D. Troug, Kohn H. Arnold, and Mark A. Rockoff who argue that patients who want to get off their ventilators should have their will honoured, but they should not be sedated *before* they are actually disconnected.[19]

This position is overly cynical. What it says is, in effect, that patients who suffer and who do not want to live any more should be *forced* to do so, unless they are brave enough deliberately to suffocate!

I think we may safely assume that well informed patients, who choose to be disconnected from their ventilators, must have good reasons to do so. There is so much evidence to the effect that, in such situations, *most* people tend rather to cling to unrealistic hopes, and request *more* treatment rather than less, that we must conclude that those who *do* give up hope and allow nature to take its course, must have very good reasons for the choice they make.

10. IS PSYCHOLOGICAL PAIN SPECIAL?

Finally we have the objection that psychological pain is special. For example, Professor Balfour Mount, who defends the position that terminal sedation can be given as a last resort, has famously argued that it should not be given in order to relieve mental suffering. As we saw, Professor Peter Strang also voiced a similar opinion in connection with the Swedish case. And, in the (highly conservative) guidelines on palliative sedation, adopted by the Norwegian Medical Association, in the aftermath of the Baerum case, it is firmly stated:

> 2. Palliative sedation should only be resorted to in exceptional cases, where an intolerable suffering is caused by, and is dominated by, physical symptoms. Psychical suffering alone is not an indication for palliative sedation.[20]

Why? The Norwegian Medical Association merely *states* its position. And so did Professor Strang. Mount, however, tries to substantiate it:

> First, our understanding of the determinants of psychological and spiritual suffering is far more primitive than our understanding of suffering that is predominantly 'physical' … How different the course of events for countless patients would have been if at that time Cicely Saunders had opted for sedation of her patients in pain! We know that patients experiencing physical pain may express a wish for euthanasia which disappears when the pain is controlled. It has been suggested that that may also be the case in the presence of psychological distress.
>
> Second, many physicians are uncomfortable in dealing with the psychological and spiritual aspects of suffering. It is probable that those individuals are less well equipped to respond helpfully when faced by these complex issues. This does not, however, preclude the possibility that others might be more effective, unless of course the patient has already been sedated into oblivion.

> Third, the presence of these symptoms does not necessarily indicate a far advanced state of physiological deterioration. Where and by whom would the line of permissible euthanasia be drawn?
>
> Finally, we are beginning to understand that quality of life, life satisfaction, and sense of emotional well-being may be unimpaired in terminal illness and that they depend not on an intact body but on the integrity of the web of relationships with self and others experienced by the sufferer.[21]

One way of understanding the first argument may be this. If we allow that patients, who want to become sedated because of mental suffering, be sedated into oblivion, then we will not learn about how to treat their kind of condition. So in the interest of science (and future patients), these patients should be forced to stay awake, in spite of their wish to be asleep (comatose).

This is not a standard view of how subjects should be enrolled in medical experiments or clinical trials. On the contrary, according to the Helsinki declaration, the research subject involved in a clinical trial should be provided the best-proven treatment for his or her condition. But the best-proven treatment is, in the cases under consideration, sedation (coma). The wish of the patient must be honoured by the doctor, then, if the doctor wants to abide by the Helsinki declaration, even if this means that some new evidence, rendering possible a better understanding of how to deal with mental suffering, will not be forthcoming. Moreover, since, for religious and other reasons, many people do not request terminal sedation, the problem to gain knowledge about psychological or spiritual suffering should not be that difficult to handle.

What are we then to say of the second argument? There is no denying that many physicians are uncomfortable in dealing with the psychological and spiritual aspects of suffering However, this is merely an argument for seeking a second opinion when a patient wants to go into oblivion. It is not a principled argument to reject the request of the patient, should the patient hold on to it, having listened carefully to the treatment offers made also by the second doctor.

The third argument could be met in either of two ways. One way would be to acknowledge only a right to terminal sedation for patients for whom, as far as we can tell, death is indeed imminent. Another would be to acknowledge also a right for terminal patients with intolerable (and incurable) psychological suffering to *make* death imminent, through sedation and treatment withdrawal. I have already indicated why I believe this is how we should treat physical pain. I see no reason to treat psychological suffering any differently.

And finally, what about the last argument? Once again, there is no denying that emotional well-being may be unimpaired in terminal illness. However, while this may be an argument for (some) terminally ill patients not to request sedation, it is no argument for patients whose emotional well-being in dying *is* severely impaired not to do so.

Many patients may have good reasons to live and endure as long as possible, or at least to the extent that they manage to defend 'the integrity of the web of relationships with self and others' that they experience. However, if the opinion of the patient is that he or she is living a life no longer worth experiencing, then it is difficult to see why his or her request for terminal sedation should not be honoured.

11. CONCLUSION

Terminal sedation comes under attack from two radically different camps. Some people who are in favour of euthanasia fear that terminal sedation will be chosen as an *alternative* to euthanasia. They know that some people who object to euthanasia accept terminal sedation. They therefore see a point in borrowing prestige from terminal sedation in their argument for euthanasia. In crucial respects, euthanasia is not different from terminal sedation, they claim. And those who are very strongly against euthanasia fear that the acceptance of terminal sedation will push us on a slippery slope, where we will soon end up accepting euthanasia as well. So they see a point in borrowing disrepute from euthanasia, in an attempt to show that terminal sedation is no better than euthanasia. Both parties concur, then, in trying to slur the line between euthanasia and terminal sedation. However, as we have seen, an important difference between terminal sedation and euthanasia exists. So it should be possible to look at terminal sedation as a kind of compromise position in the discussion about euthanasia.

Adherents of euthanasia may well argue that terminal sedation is not good enough. Some patients may want to be intentionally and actively killed by their doctors, they may claim. However, while they continue to argue their case, they should be prepared to admit that terminal sedation renders dying easier for the very patients on behalf of whom they put forward their argument for euthanasia.

And adherents of the Sanctity-of-Life Doctrine, who oppose euthanasia, should be able to appreciate that there exists a way for them to answer the stricture that they are insensitive to human suffering. They can accept a practice of terminal sedation and yet, for all that, stick to the Sanctity-of-Life Doctrine and their opposition to euthanasia.

Torbjörn Tännsjö
Department of Philosophy
Stockholm University

NOTES

[1] The example is presented in, and is followed by, a discussion in several issues of the journal. See also the selective bibliography of the literature since 1990.
[2] Gillian M. Craig, 'On withholding nutrition and hydration in the terminally ill: has palliative medicine gone too far?', *Journal of medical ethics*, Vol. 20, 1994: 139-143.
[3] R.J. Dunlop, J.E. Ellershaw, M.J. Baines, N. Sykes, and C.M. Saunders, 'On withholding nutrition and hydration in the terminally ill: has palliative medicine gone too far? A reply', *Journal of Medical Ethics,* Vol. 21, 1995: 141-143.
[4] Lawrence J. Schneiderman, 'Is it morally justifiable *not* to sedate this patient before ventilator withdrawal', *The Journal of Clinical Ethics*, Vol.2, 1991: 129-130.
[5] Many philosophers have attempted to make this kind of distinction, most ingenious among them all, Jonathan Bennett. But in my opinion even Bennett fails. Those who want to check this assessment of mine should read his most recent treatment of the subject, which can be found in his book, *The Act Itself* (Oxford: Oxford UP, 1995), chapters 4 through 8.
[6] An authoritative statement of the principle of double effect can be found in *New Catholic Encyclopedia,* McGraw-Hill, New York, 1967.
[7] Airedale NHS *Trust v. Bland* (1993).
[8] 'The Additive Fallacy', *Ethics*, 1988:5-31
[9] American Association for Hospice and Palliative Medicine, Position statement on the end-of-life, Appendix of this volume, p. 128.
[10] 'Palliative Options of Last Resort', *JAMA*, Vol. 278, 1997. In this book reprinted as Chapter 1, p. 6.
[11] Ibid., p. 7.
[12] Ibid., pp. 7-8.
[13] 'Is Sedation Without Hydration or Nourishment in Terminal Care Lawful?', *Medico-Legal Journal*, Vol. 62, 1994:200.
[14] Ibid.
[15] At a meeting held at The National Board of Health and Welfare on Oct. 29, 1997.
[16] Håkan Samuelsson and Eva Thornberg, 'Symptomlindring inte dödshjälp' (Symptom control is not euthanasia), *Göteborgs-Posten*, January 1, 1998.
[17] This description it taken from M. O'Connor, D.W. Kissane, and O. Spruyt, 'Sedation of the terminally ill – a clinical perspective', *Monash Bioethics Review,* Vol. 18, 1999, p. 22.
[18] Dag Lundberg, 'Psykologiska och etiska aspekter' (Psychological and Ethical Aspects), in Matts Halldin and Sten G. Lindahl (eds.), *Anestesi* (Anaesthesia) (Stockholm: Liber, 1999), p 190. My own translation.
[19] 'Sedation Before Ventilator Withdrawal: Medical and Ethical Considerations', *The Journal of Clinical Ethics*, Vol. 2, 1991:127.
[20] Guidelines on palliative sedation adopted by the Norwegian Medical Association, Appendix of this volume, p. 132-133.
[21] 'Morphine Drips, Terminal Sedation, and Slow Euthanasia: Definitions and Facts, Not Anecdotes', *Journal of Palliative Care,* Vol. 12, 1996:34.

CHAPTER 3

MAGNA ANDREEN SACHS

SEDATION – UNCONSCIOUSNESS – ANAESTHESIA! WHAT ARE WE TALKING ABOUT?

I. INTRODUCTION

In this chapter I will start off by briefly dwelling on my mission as a doctor and especially as a caregiver involved in end-of-life care. I will then expand on the subject 'sedation' and its place in end-of-life care – in which it is called 'terminal sedation' – and make sure that we all know what we are talking about and aiming for: sedation or unconsciousness, maybe even anaesthesia. And so we call a spade a spade - and a hammer a hammer – and do not mix up the names of the tools. They are both very useful tools, as are sedation, unconsciousness and anaesthesia – but not for the same purposes! I will then finish off by briefly touching upon the vast difference – in my opinion – between 'not preventing death from happening' and 'actively put an end to life'.

2. PALLIATIVE CARE

As a doctor my mission is to help patients/clients live well even with a serious disease. This mission becomes even more demanding when the person is dying, be it from dementia, stroke, chronic heart disease, chronic obstructive pulmonary disease or any other fatal disease which slowly kills the person who is stricken by it.

Somewhere along the course of these illnesses it becomes evident that there is only one end to this course. There is no treatment that can change the fact that this person is going to die from this disease – or rather these diseases – because very often the patient has several fatal diseases at the same time. That is the moment at which you change the focus of your care for the patient from treating illness to end-of-life care, i.e. palliative care. Cure turns into care. But my mission as a doctor does not change there. It is still: to help the patient – as best I can – to live well, whatever time is left (days, months, years), and to die in dignity.

Torbjörn Tännsjö (ed.), Terminal Sedation: Euthanasia in Disguise?, 31-35.

Palliative care entails attention to *physical, psychological, social* and *spiritual* needs. Those are the four pillars of palliative care. The aim of palliative care is: quality of life, i.e. a meaningful life until the end.

Relief of *physical* symptoms is a very central part of palliative care. Those symptoms are pain, nausea, dyspnoea, constipation and fatigue just to name the most common ones. Today there are excellent strategies for coping with all of them.

Psychological symptoms are common – especially when physical symptoms are not satisfactorily addressed. They range from hopelessness and fear to depression. Very often the fact that the patient is depressed is overlooked. Caregivers tend to regard a depressed mood as being natural and appropriate, considering the circumstances. By doing so and by not paying attention to the possibility of depression the patient may not have the benefit of effective antidepressant drugs combined with psychotherapy and/or psychostimulant drugs.

One sign of depression is the desire to die. This can be a result of psychological, social and spiritual needs not being recognized and attended to early in the course of the illness. Maybe there are also unattended physical symptoms. A wish to die should evoke active attention to those needs. It is a cry for help – an S.O.S.

Spiritual support is an essential element of palliative care. Questions about spiritual needs, beliefs, thoughts and wishes should be asked early in the phase of end-of-life care in order to make sure that all necessary steps are taken to achieve quality of life at the end of life. 'Advance care planning' is part of the strategy to provide spiritual support as well as social support – the fourth pillar of palliative care.

Social support includes caring for the family's needs while at the same time attending to the dying person's worries concerning practical and/or financial matters that might be a burden to the family after death.

All steps undertaken to alleviate symptoms should be evaluated by the use of *assessment scales* to confirm that actions taken are effective and goals are met.

The end point of palliative care including all 4 pillars is, as has been stated: *quality of life*. This, too, can and should be measured. There is no other way to ascertain whether or not we are doing the right things at the level of the individual as well as the population.

3. SEDATION

Sedation may be an active part of palliative care, especially if the dying person suffers from delirium in the terminal stage. The use of sedating drugs does not automatically imply sedation. Sedatives can produce a range of effects. Sedation is one *effect* – among others – that can be achieved using sedatives. It is always the effect we are aiming for. Sedation denotes a level of consciousness at which you are comfortable – yet open to contact. A patient with delirium is restless, filled with anxiety, disoriented and closed to communication. It is a dangerous condition and it is painful and calls for treatment. In that situation you aim for sedation.

But before starting to administer sedatives for this purpose: you have to exclude reasons other than sensory dysfunction! There may be somatic reasons for delirium, like sepsis, constipation, physical pain, dehydration, or metabolic disorder, and if so: each condition should be specifically treated. Note that sedative drugs are not analgesic drugs and should never be used for the purpose of treating pain. Another reason for a delirious state may be inappropriate medication. Medications need to be reviewed: which drug or combination of drugs might cause delirium? The way the person is looked after has to be reviewed as well: is there something in the care setting that is frightening or causing pain?

If all treatable causes of delirium have been considered and the patient is still restless and agitated, treatment with sedating drugs in order to achieve sedation is justifiable.

Now, if sedation is the aim and your intention is to use sedatives you must verify that the aim is met. This is done by way of an assessment scale. Why? Because sedation does not come automatically and does not remain stable simply by administering a sedative. Managing sedation calls for close monitoring and observation of the patient. If patients are not monitored, they are likely to become overly sedated, i.e. unconscious and even fully anaesthetized.

And the reason? All sedatives are central nervous depressants, i.e. potentially anaesthetic drugs and thus also potentially lethal drugs. Their therapeutic span ranges from light sedation through unconsciousness, to various anaesthetic levels, and even to the cessation of vital functions, i.e. death. To lie for hours and days in an unconscious almost anaesthetic condition is in and of itself life-threatening and eventually lethal. During surgery and in the intensive care unit, complications due to unconsciousness and the anaesthetic state are avoided by continuous monitoring of circulatory and respiratory parameters and intensive circulatory, ventilatory and general care.

This is a question of dosage, or more precisely: of blood concentration. Which blood concentration will cause what degree of depression cannot be predicted and depends on characteristics specific to the individual. Furthermore, which dosage will cause what blood concentration is also a matter of individual variation. So, you really see that you cannot predict an effect by prescribing a certain dosage. That is why it is recommended to begin with small repeated bolus doses and monitor the effect carefully, and in the event administration more than once an hour is required in order to maintain desirable effect levels, then for the sake of efficiency you could start an infusion while closely monitoring the degree – or level – of depression by the use of an assessment scale, adjusting the infusion rate according to the results obtained.

The level that is aimed for and that is called 'sedation' is the level at which the patient is still communicating and responsive, but calm and cooperative. Further down the scale the patient is 'wakeful' i.e. asleep but responsive to touch and commands. If blood concentration rises still further the patient becomes unconscious with reactions only to very distinct and painful stimuli. This is close to the 'anaesthetic state' and calls for anaesthetic care. On the far end of this scale – before death – is 'no reaction even to painful stimuli'. That is 'anaesthesia'! Anaesthetic care is called for here in order to avoid immediate or delayed death from underventilation, loss of vital functions, thromboembolism etc. The aim of anaesthesia is to prevent any awareness of pain and other sensations and to provide optimum conditions for the surgeon. This is in accordance with the interest of the patient and is appreciated by him/her when fully awake after anaesthesia – i.e. after discontinuation of anaesthesia and surgery.

You cannot appreciate anaesthesia when anaesthetized. You have to be conscious to be able to value what you are experiencing, to sense the loving care of your family, to appreciate comfort and relief of symptoms. Quality of life cannot be sensed in the unconscious state.

Thus, in all you do and especially in palliative care the AIM should always be clear. The decision of course has to be made in consultation with and agreement with the patient. If the patient is incapable of participating in such decision-making it is essential to have a good picture of what the patient would have wanted had he been able to express his will. The aim should be clearly established with the staff and the family so that they too know: why are we doing this? What is it that we are trying to accomplish that might add value to life or reduce suffering during the few hours, days, weeks or months that are left? And how do we know that these aims are met? By asking the dying person herself and/or by using assessment tools for evaluation.

4. DYING AND KILLING

For me there is a clear distinction between dying and killing. As a doctor you do things that have an effect on the process of dying: you administer chemotherapy, you give parenteral nutrition, etc. You can also choose not do these things – i.e. let the dying process run naturally without intervention. Or you can choose to stop doing things that you have initiated earlier because it is so obvious that it causes more suffering than it adds value to the life that is left. In any case your prime task as doctor and caregiver is to alleviate symptoms, prevent suffering and promote meaningfulness. All these decisions are of course made in consultation with the patient or the assumed will of the patient. These actions affect the natural process of dying as a part of life. And they are undertaken to serve the life by ensuring that quality of life as appreciated by the individual person is present to the last breath.

To introduce an agent that is separated from the dying process – like a lethal drug in lethal dosages – and thereby end the life of that person is *killing*. I do not kill – I help people live well even when they are dying.

5. CONCLUSION

In conclusion, I have described the concept of sedation as it is defined within medicine. Professor Tännsjö has another definition of the word 'sedation', and for the sake of clarity I think it is essential that we share the same understanding of the word.

Sedation – whether terminal or applied any time during life in association with critical illness – is not euthanasia in disguise, because in its correct sense and correctly applied it is not lethal. Sedation is a part of palliative care. Sedation – like any other treatment – should be evaluated and used on the correct indication and in concert with the patient's will or assumed will.

Let me finish off by stating once again: my professional mission as a doctor is to serve life and to decrease the burden of illness. It is not to extinguish life. It would be a serious encroachment on my professional integrity to require any such thing from me.

Magna Andreen Sachs
Karolinska Institute
Stockholm

CHAPTER 4

GUNNAR ECKERDAL

SEDATION IN PALLIATIVE CARE – THE DOCTOR'S PERSPECTIVE

1. INTRODUCTION

In this chapter I will recapitulate three cases where I have treated patients with very different diagnoses. One of the patients had a serious but rather trivial disease. The patient had her request for antibiotics rejected. Two of the patients were terminally ill. In one of these cases the patient was 'terminally sedated', in the sense defined by Professor Torbjörn Tännsjö in the opening chapter of this book. In a second case the demand for terminal sedation was rejected.

Before embarking on the description of these cases, however, I will start by giving a rough characterisation of how I conceive of the role of the doctor and of the patient. This role is the same, basically, irrespective of whether we are speaking of difficult end of life decisions, or of the cure of more mundane diseases.

The patient is the one who chooses to go to the doctor. Doctors on the other hand are expected to make a diagnosis, to inform the patient about the diagnosis in terms that are comprehensible to the patient, to offer treatment, and to support the patient. Moreover, the doctor is supposed to follow up how the patient is affected by the treatment. Doctors are educated and trained to do such things. And a crucial requirement facing the doctor is that his or her work should be based upon clear evidence.

The goal of medical practice is twofold: to provide longer life, and better quality of life, for the patient. In palliative care, however, the latter aspect of the goal is in focus. It should also be noted that the two aspects of the goal may come into conflict with each other.

Doctors are supposed to offer good, evidence-based medical practice; this is required of the doctor also in the case where the patient is dying. The doctor must never lie about the facts of the case. The patient should be informed that there is always much the doctor can do, but the patient must also be informed about the fact

Torbjörn Tännsjö (ed.), Terminal Sedation: Euthanasia in Disguise?, 37-41.

that the doctor cannot do everything. The patient should in particular be informed about side effects of a suggested treatment, and the patient should be informed of the sad fact that, in some cases, the patient will die, whatever the doctor does.

The patient should be informed that there are potential positive effects of each treatment as well as potential negative side effects. The doctor has to consider the potential effects/side effects for the patient, the relatives and other patients. In many cases the doctor must consult colleagues.

The doctor has the exclusive responsibility of weighing costs against benefits of each possible treatment. Before a decision about treatment is reached the patient should be fully informed about effects and side effects, and given the right to reject the suggested treatment. If, from the point of view of the cost/benefit-analysis, two alternative treatments are equivalent, then the wish of the patient is of the utmost importance. However, a patient has no right to 'require' that a certain treatment be provided. When in doubt about a treatment requested by a patient, the doctor him- or herself has to reach a decision. This is true in general, and it is no less true when it comes to terminal sedation.

2. FIRST EXAMPLE: BARBARA BORN –83

The patient, Barbara, born 1983, calls her GP. The patient has had a slight temperature for two days, she coughs, and she is suffering from pain in her ears. The doctor examines her and detects a virus infection. Her anamnesis is taken, and it transpires that, during the years the patient has had 10 otitis episodes, which have caused a slight hearing deficit in her right ear. The doctor realises that antibiotics can reduce the risk of deafness. However, the cure may well be useless (if, in fact, she is only suffering from a virus infection). On the other hand, the cure may just as well turn out to be harmless (if no side effects occur). However, if side effects occur, the cure may come to cause lethal allergic reactions. The antibiotics cure can moreover increase the problem of resistance to antibiotics in society. So there is no denying that there are both pros and cons in relation to the possible antibiotics cure. The GP has to weigh the pros and cons, to inform the patient, to listen to her wishes for antibiotics and then to make a decision. In the circumstances a decision was reached: No antibiotic treatment was provided. The decision was taken by the doctor, who had to take sole and full responsibility for it.

3. SECOND EXAMPLE ANN BORN –76

Ann suffers from a brain tumour. She is unable too see, talk, or hear. She is capable of using her right arm, however. She is in no way communicable. While awake she persistently and aggressively hurts herself with her arm. She is often shouting, and she is showing obvious signs of suffering. The very last days her symptoms have progressed. No curative treatment whatever is possible. The surgeon asks the doctor in the palliative support team for advice. The doctors discuss the case together and in the team as well as with the parents of the young girl. It is agreed that the patient suffers. Her diagnosis is considered certain. There is no possibility of providing her any quality of life in the future. It is decided that permanent induced sleep ('terminal sedation', as defined in the opening chapter of this book) is the best palliative option. The patient is put to sleep. Hydration and/or nutrition are withheld, not as a means of shortening her life but on the contrary because, as a matter of fact, these measures would shorten her life (by worsening the oedema of her brain). Midazolam is given as her sole treatment. The patient dies after 48 hours of induced sleep (coma).

4. THIRD EXAMPLE: GEORGE BORN –47

The patient, George, born in 1947, has lung cancer. He is very tired, his pain is treated effectively, and he suffers from no dyspnea. However, the patient is often showing signs of anxiety. The patient refuses to discuss the background of his anxiety. Intermittent sedation is started. Midazolam is used more often, but the patient asks for permanent treatment. He claims that he does not like the periods of enforced consciousness. The patient is provided with higher and higher doses of Midazolam, some days as much as 10 times/day. One day, between two doses, he makes the following statement:

'OK, I give up! I realise that I'm dying. Please let me talk to my sons!'

The sons of the patient are informed of their father's wish, they enter the ward and they talk to their father. The patient deteriorates and dies after 6 days. During this period he asks for Midazolam only 1–2 times every night.

Note that it was the doctor's decision not to provide terminal sedation. As the case turned out, this seems to have been the right decision. It allowed the patient/father to reconsider his situation and to see his two sons.

5. TERMINAL SEDATION AT REQUEST?

We have seen that there are examples where terminal sedation can be given. Does this mean that terminal sedation should be given at the patient's request? In my opinion, terminal sedation cannot be given at the patient's request. There is no right to terminal sedation. Terminal sedation cannot be ordered by a patient.

The doctor must fight for his patient's autonomy, and, when necessary, be 'paternalistic' (to act in the best interest of the patient). And in particular, if the patient asks for help to die or to sleep for the rest of his/her life, the doctor must act in a way that allows, if possible, for the patient to change his/her mind. If the doctor on medical ground sees a potential of recovering, he must take that into account when choosing treatment.

Autonomy is a value in the medical context. However, it is not only of importance for a physician to *respect* the autonomous decision of patient. Sometimes the physician has to *guard* the autonomy of his or her patient. Sometimes the physician must act as a guardian of the autonomy of the patient, against the patient him- or herself. A person who can never again become awake, a patient who is sedated into oblivion, is robbed of his or her autonomy. This is not acceptable. No doctor should be required to rob a patient of his or her autonomy.

Faced with a conflict where, by respecting the wish of a patient, the doctor will rob the patient of his or her autonomy, it may be necessary for the doctor to reject the demand from the patient – in the interest of the autonomy of this very patient.

Total autonomy can never be achieved So while terminal sedation is a part of good palliative care, which can be provided when a patient has lost his autonomy, in order to save the patient from suffering, it cannot be provided at the request of a competent patient. No doctor should accede to such a demand from a patient.

We sometimes hear that, unless a patient knows that his or her wishes will be the doctor's demand, the patient will not feel safe in relation to the health care system.[1] Because of the lack of trust, the patient will even be reluctant to seek medical advice in the first place. There may be some truth to this saying. But the truth in it is that the doctor should always respect a wish from a competent patient not to be treated. The patient has an absolute veto against all sorts of treatment. However, this does not mean that the patient should have a right to request whatever treatment may be available, such as terminal sedation. When we are young and healthy, we can easily understand that, when sick and helpless, we are in many cases likely to succumb to all kinds of irrationality. So we can accept that, in those circumstances, we need doctors upon whom we can trust. We need them as guardians of our autonomy. The existence of such doctors is no reason, at least not to most people, not to seek medical advice. Quite the contrary, the existence of such doctors is a reason for most people to feel secure in relation to the health care system.

Good palliative medicine comes to the same thing as good evidence-based medical practice. Dying patients should be treated just as well as patients for whom a cure is available. But this may mean, in some circumstances, that the wish of the patient is not respected. In some circumstances, the patient must be guarded against his or her own decisions – in order to safeguard the autonomy of the patient.

We need no special regulations for dying patients. This is true also with respect to anaesthesia. In many situations patients are being put to sleep. The duration of the treatment is as short as possible, both because there is no need for long periods of induced coma, and because there are definite risks associated with keeping the patient comatose. When the treatment starts, and the patient is put to sleep, the treatment has a definite goal. It is crucial that the treatment is carefully assessed. The patient must be carefully monitored. Every day one or more new decisions are made regarding whether or not to continue the treatment.

In some cases the treatment (the sedation) continues until the patient dies. This is the best decision if it is obvious that the advantages associated with the treatment are greater than the risks. The doctor shares the weighing of risks and benefits with the team, the patient and the relatives. However, in the final analysis, the doctor on his or her own must take full responsibility for the decision.

So where do the philosophers enter the picture? Doctors need philosophers. Patients need doctors who from time to time listen to philosophers.

Gunnar Eckerdal
Bräcke Hospice
Gothenburg

NOTES

[1] Cf Torbjörn Tännsjö, *Coercive Care: The Ethics of Choice in Health and Medicine* (London and New York: Routledge, 1999) for a defence of this view.

CHAPTER 5

SIMON WOODS

TERMINAL SEDATION: A NURSING PERSPECTIVE

1. INTRODUCTION

Terminal sedation raises a number of questions and problems not the least of these is that of the definition of terminal sedation. In chapter 2 of this volume Professor Torbjörn Tännsjö defines terminal sedation as 'a procedure where through heavy sedation a terminally ill patient is put into a state of coma, where the intention of the doctor is that the patient should stay comatose until he or she is dead'. If Professor Tännsjö's claim that terminal sedation can be distinguished from euthanasia and moreover, that terminal sedation is an alternative to euthanasia, is valid then it would seem that terminal sedation is an intervention potentially compatible with good nursing care at the end of life. However other contributors to this volume argue that terminal sedation is nothing other than a slow form of euthanasia. Terminal sedation also raises questions concerning the nature of the good death and what constitutes good care at the end of life. In this chapter I will say something about the history of nursing in relation to care of the dying, articulate some of the goals of good nursing care of the dying and say in what ways the use of sedation in the care of the dying is a valid intervention compatible with the goals of nursing care. In addition I will draw a more careful definition, that of *palliative* sedation which has the advantage of including Professor Tännsjö's definition of terminal sedation but is more sensitive to the spectrum of circumstances in which sedation is justified. I will also set out my reasons for disagreeing with the inclusion of decisions regarding hydration and nutrition withdrawal in this definition.

I write this piece as a former cancer nurse with ten years of clinical experience, as a philosopher working in the field of health care ethics, and from the perspective of English common law; on the assumption this is consistent with many other common law jurisdictions.

Torbjörn Tännsjö (ed.), Terminal Sedation: Euthanasia in Disguise?, 43-56.

2. NURSING CARE AT THE END OF LIFE

The modern palliative care movement has its origins in a number pioneering events and individuals. In the UK palliative care has been strongly influenced by the early hospice movement. Hospice philosophy was significantly progressed from the 1950's onwards by Cicely Saunders amongst others.[1] Working and writing as a nurse Cicely Saunders was one of the first scholars to set out the basic principles of the care for the dying: 'Care for the dying person should be directed no longer towards his cure, rehabilitation or even palliation but primarily at his comfort.' [2] It was the emphasis on comfort and care that was the inspiration for many nurses to work with the dying whom they believed had been abandoned by mainstream healthcare. Care of the dying, until very recently was predominantly a nursing domain with hospices being run by nurses calling on the services of a local General Practitioner. The now international recognition of palliative medicine as a medical speciality has led many to claim that medicine has colonised and medicalised the care of the dying to the detriment of terminal care.[3] However the influence of nursing on palliative care is evident in the universally accepted approach to palliative care: a team approach, providing holistic care which includes attention to physical symptoms, psychological and spiritual issues, and incorporates the patient's family and close ones.

However one of the changes, and sources of tension, within modern palliative care is the greater readiness to resort to more aggressive forms of treatment. Such treatments would, from a hospice perspective, have been regarded as *extraordinary*, and therefore not indicated. Contemporary palliative care sees the use of such interventions as justified because it regards itself as a more inclusive discipline, aiming to treat people with active progressive disease who may be a long way from the terminal phase of their illness. The emphasis is still upon the control of symptoms and improving quality of life rather than curing the primary disease but the boundaries around what is appropriate or inappropriate treatment has become much less distinct.[4]

For nurses the emphasis is still very much on *holistic* care this includes physical care of the patient, psychosocial support of the patient and family and a greater willingness to use complementary therapies to enhance the comfort of the patient. The emphasis on comfort and the relief of suffering clearly gives priority to the subjective states of the patient in determining what the priorities of care are. This emphasis in turn influences the ethical underpinnings of practice, which is patient centred and premised upon a principle of respect for persons. An important way in which this principle is cashed out is the priority given not only to the person's report on their level of suffering or comfort but to respecting the wishes of the person for how they want to live their life to its close. With these thoughts in mind I turn now to consider what role sedation might play in palliative care and in particular how we might draw up an account of what constitutes appropriate and inappropriate use of sedation.

3. ARE SOME EXPERIENCES NOT WORTH HAVING?

Sedation is usually indicated when a person is anxious, agitated or pathologically restless. The purpose of sedation is to bring about a state of relaxation both physical and mental, to bring about sleep and to subdue awareness of unpleasant experiences. There are many circumstances in which sedation, as a medical intervention, is indicated. Consider the case of the patient who is being intensively treated for cancer. Treatments that combine drug and radiation therapies, although potentially curative are very arduous for the patient and some treatments cause patients to experience a period of severe nausea often combined with vomiting which may last several days or longer despite the use of modern anti-emetic drugs. Patients who have such a severe reaction become exhausted and psychologically very low and in such cases a regime of sedation may be indicated using a drug such as Lorazepam. Lorazepam is a benzodiazapine used to treat acute anxiety states. Administered intravenously it has a strong sedative effect and most people on this drug will sleep or experience a somnolent state for long periods, and although they may continue to vomit during this time they usually benefit from an amnesia induced by the drug, resulting in a few 'lost' days during the worst of the nausea and vomiting. No one reading this would consider such a loss to be of any significance, indeed I would suggest that most people would consider such a loss to be a positive benefit to the patient.

This example rings true with at least *one* philosophical account of the good life, namely hedonism. Hedonism is the view that the goodness of a life is contingent upon the quality of the experiences a person has during that life. On the hedonist view a good life is one in which pleasurable experiences outweigh unpleasant ones. Whether one is inclined to agree or disagree that hedonism is the whole truth with regard to theories of the good life there is no escaping from the fact that one's capacity for living the good life is severely restricted if every waking moment is overshadowed by an acutely unpleasant experience, unremitting pain, agitation, nausea and vomiting, or breathlessness. As I have said a person is no worse off, indeed, intuitively one would consider her better off for not having to undergo, or at least *remember* undergoing the experience of nausea and vomiting. This is not to say that a person could not find some meaning in the experience, for example she may consider the ardour of cancer treatment to be a personally beneficial experience, knowing that she has fought and won against a killer disease. Some people are able to find strength in the fact that they have coped with a serious illness and are able to bring this strength to bear on other aspects of their life. The possible meaning that people find in their experiences of coping with challenging situations is almost limitless. However whatever the meaning this experience may have for a person it is the meaning *she* and she alone finds in the experience that endows it with value. This is not to say that others cannot be inspired vicariously by the fortitude and courage of others; we can and are. However the benefit is strictly parasitic upon the value with which the person whose experience it is imbues the experience. It would seem improbable, or at least sadistic, to claim that there is virtue in the suffering of a person if they found no meaning or value in the experience themselves. If this were not the case then a powerful motive for relieving the suffering of others would be removed. An important principle in the modern management of pain and other symptoms is that the person who says they have pain should be believed and their own testimony should be the main reference point when adjusting their pain relief. This principle is crucial to understanding what constitutes good nursing or medical care of people who are suffering pain or other symptoms.

In addition to the relief of suffering there are also other virtues at play in caring for a suffering person. These virtues are seldom described in any detail in the ethics literature perhaps because they fall within the general category of beneficence, the usual level of abstraction at which these matters are discussed. However describing these is an opportunity to elaborate on what constitutes good nursing care. It is an assumption in palliative nursing that good nursing care may contribute to rendering a challenging experience, if not meaningful, then more bearable.

Consider the patient who has received sedation for her severe reaction to treatment. Despite the administration of sedation at regular intervals she is likely to continue to experience some nausea and will vomit. She may sleep for long periods or else be drowsy but rousable. Consider what a nurse might do in caring for her during the night when her symptoms arc at their peak. The nurse will observe her at regular intervals of 15 or 20 minutes, if necessary going to her bedside, quietly, discretely and without switching on the light. The nurse will remove any receptacles containing vomit, replacing these with sufficient clean ones; if she is awake she/he will enquire after her needs, offer a mouthwash, a glass of water or perhaps some ice to suck. If necessary she/he will change her bedding and clothing, stay with her if she is retching, perhaps, if appropriate placing a comforting hand on her back. Sometimes during the dark small hours of the morning a patient like this will wish to talk, speaking of her hopes and fears and feelings. Often the nurse need do nothing more than listen attentively, and many patients have said what a benefit it is to have someone to listen to them or simply to be with them at such times. I have included this account of nursing care, a composite anecdote from my own experience as a cancer nurse, as an explicit example of the virtue of caring. There is of course research evidence which supports the benefits of such caring but I have chosen not to site this as I believe most people will respond intuitively to this account of the care given by one human being to another as an obviously good thing. This is also to remind the reader that sedating a patient even to a point at which they are deeply asleep or comatose is not to abandon the person since even if this patient were to be deeply sedated they would still be cared for with respect and dignity.

In this example I have used an account of a person receiving intensive and potentially curable treatment. However would our intuitions differ if the lost days were at the end of life rather than in the middle? Although I have argued that it is the person whose life it is who gives meaning to their life, even out of their suffering, it would seem more plausible that this would occur in moments of reflection at a time after the period of suffering. This is of course not an opportunity the dying person has. No doubt some people will wish to experience their life to its end, perhaps because they wish to cling to life itself or because they have a particular religious belief about the end of life, however, there can be no obligation to experience one's life to its end and in many cases not doing so can be a positive benefit in the same way that avoiding the experience of nausea is a benefit.

Having given an example of how sedation may be appropriately used and combined with good nursing care I now turn to the question of terminal sedation.

4. WHAT IS TERMINAL SEDATION?

In his introduction to this volume Professor Torbjörn Tännsjö defines terminal sedation as 'a procedure where through heavy sedation a terminally ill patient is put into a state of coma, where the intention of the doctor is that the patient should stay comatose until he or she is dead'. Professor Tännsjö's position is that terminal sedation can be distinguished from euthanasia and furthermore is an ethical alternative to euthanasia in countries where the general ethos is against euthanasia. I believe that such a distinction *is* feasible and that sedation ought to be considered a valid option at the end of life. I do not agree that 'terminal' sedation is an appropriate description of this intervention, and I believe that the distinction between sedation and euthanasia is much harder to sustain when other decisions, for example to withhold hydration and nutrition are not considered as distinct decisions in their own right. For the record I am also broadly sympathetic to arguments in favour of euthanasia but see that attempts to reconcile killing and caring in a nursing context are highly problematic.

Professor Tännsjö also identifies three positions with regard to terminal sedation (1) that it should never take place, (2) that it should take place, but only as a last resort, and (3) that it should be considered a normal part of palliative care and that a doctor should be free to provide it at the terminally ill patient's request. Like Professor Tännsjö I find the third position defensible but would also consider the second position as the one most compatible with nursing approach to care at the end of life.

5. WHAT IS IN A WORD?

Professor Bert Broeckaert has argued that the use of sedation in the context of palliative care is a valid palliative intervention.[5] He further argues that referring to the use of *terminal* sedation is ambiguous, misleading as to the nature and purpose of sedation and does not sufficiently distinguish the practice from forms of euthanasia. Professor Broeckaert's positive account of 'palliative sedation' covers some useful ground, and helps to show how sedation is compatible with good nursing care in the palliative context. The arguments also suggest ways in which sedation in the palliative context can be distinguished from euthanasia.

Professor Broeckaert argues using an approach generally used within the palliative literature, that the use of the term 'terminal' is unnecessarily negative in its connotation. This is a familiar point from within the palliative care community, which over the past 40 years has attempted to define itself both as a new health specialty and as a speciality that makes an active contribution to the care of those with progressive disease. In the UK context, a country, which is one of the pioneers of palliative care, there was a deliberate decision to adopt the term 'palliative' care in place of terms such as 'hospice' or 'terminal' care. The historical and conceptual intricacies of this debate are too removed from the present discussion but suffice to say that the concept of palliative care has been deliberately manufactured so as to convey something positive about the care of people with incurable disease from their diagnosis through to their death.[6] This positive account is reflected in the WHO definition of palliative care and is to be found reflected in the many national statements of palliative care philosophy adopted by the numerous countries that have made a commitment to palliative care. [7]

In addition to its negative connotations 'terminal sedation' is also ambiguous on several fronts, semantically and conceptually. For one thing a common understanding of 'terminal' is that it is synonymous with the terminus or end, and in this case, death. Palliative care however is not solely concerned with the terminal stages of a disease but sees itself as engaged across the spectrum of active disease, which in many cases may be years from death. If sedation is an appropriate intervention for symptom relief then what should the use of sedation in the non-terminal phase of care be called, 'non-terminal sedation'? The point is that any prefix seems redundant unless it contributes something meaningful to the term to which it is attached. This in turn raises the question as to the justification of using the term 'terminal sedation' in any context. Professor Broeckaert suggests that one reason not to use this expression is because it falsely implies a more active causal role for the sedation itself, which is suggestive of some form of euthanasia, a highly undesirable implication from the palliative care perspective. For one thing, using a term for a practice that carries such implications may disadvantage those who may stand to benefit, for example patients who most likely would benefit from sedation might refuse to consent for fear that this was a form of euthanasia. Professor Broekaert suggests that a term is needed for this practice, which is clear both about the criteria for use and the intention behind the intervention; he suggests the term 'palliative sedation'.

At first glance it is not clear that the term 'palliative' is itself a distinct improvement on other candidate terms. However without becoming distracted with the details of the history of this term it is quite clear that a consensus has emerged regarding palliative forms of treatment within the broader context of palliative care. An early impetus for the development of modern palliative care was concern for the poor quality of care offered to dying people. Early work within the hospice setting in the UK, under the influence of such people as Cicely Saunders, concentrated on good symptom control so as to improve the quality of life of people with incurable disease. From an early stage there was the intention to achieve a balance or therapeutic ratio between the burden of medical interventions against the severity of symptoms and improvements in quality of life. One of the elements of this therapeutic ratio was the dilemma of under-treatment versus the problems of iatrogenesis. At this stage in the 1950's and 1960's little was still known about the phenomenon of pain and the ways in which it could be treated. Although morphine was readily available its mechanism was not fully understood and many doctors were in such fear of causing side effects that when morphine was used it was used in inadequate ways. The side effects that doctors were concerned about included causing addiction, but more serious was the fear of causing premature death, because morphine may cause a fatal depression of the patient's breathing. However, on this latter point a different approach was adopted by some doctors who argued that when there was no other way of relieving a patient's suffering then side effects, even fatal side effects are justified. It is here that the familiar justification for such practices, touched on by Professor Tännsjö in chapter 2 of this volume, comes into play. The argument is that treating a patient in such a way that the treatment shortens his or her life can be distinguished from murder or mercy killing because of the intention of the doctor to treat rather than kill, and because the treatment used has more than one effect. This enables the doctor to *intend* the beneficial effect and *foresee* but not intend the negative effect. This very point was made by Lord Devlin in a now famous legal case, R v Adams [1957] where he commented: 'If the first purpose of medicine, the restoration of health, can no longer be achieved, there is still much for the doctor to do, and he is entitled to do all that is proper and necessary to relieve pain and suffering even if the measures he takes might incidentally shorten life.' [8]

This introduction into English law of the 'doctrine of double effect' (DDE) has been the cause of much debate, and as the discussions in this volume show, are directly relevant to the issue of terminal sedation. It is necessary therefore to spend some time on this distinction but for the moment I shall do so in relation to the treatment of pain. I, like Professor Tännsjö, believe that there is a plausible defence of the DDE and I shall argue for this in the context of palliative care.

One of the central criticisms of the DDE is made by those who see the consequences of an act as the determinant of the moral character of the act. Hence when two acts lead to equivalent outcomes they are therefore *morally* equivalent acts. This is the point made by Helga Kuhse and others when they say that bringing about a person's death by an act, which foresees but does not intend their death as a consequence, is morally equivalent to directly and intentionally bringing about their death. I shall say nothing further on this point other than to state that both intentions and consequences in my view are relevant to the moral nature of an act. To say anything further on this issue would be to repeat what Professor Tännsjö has already said in his useful discussion. Instead I shall discuss a second criticism of the DDE, which concentrates on the role ascribed by its supporters to intention. Critics of the DDE argue on two grounds, one is that it is difficult, if not impossible to distinguish intending from merely foreseeing an effect. Secondly, because intentions are subjective states there is the problem of verifying just what the intention of the actor was in a given situation.

Whilst it may be difficult and sometimes impossible to distinguish intending from foreseeing it is surely not the case that it is always *impossible* to judge another's intentions. If it were so it would be impossible to convict in the criminal courts, beyond all reasonable doubt, that the accused was indeed guilty. To judge what another intended relies not only upon first person testimony, an insight, if the person is truthful, into the *subjective* conditions of their intention, but also on the *objective* conditions, the facts of the matter, what they did and how they acted. Of course this is not to say that this is an infallible approach, indeed there are uses of the DDE in which the minimal or sufficient conditions for rendering an act 'innocent' are met but in such a way as to obscure the real intentions at play. I refer to this as the 'Cynical Doctrine'. The legal case R v Adams [1957] is perhaps one such example where the doctor (Adams) was acquitted of deliberately ending the lives of his patients by increasing the dosage of opiates administered to them, in circumstances in which he stood to inherit money from his patients. Adams was acquitted because his actions met certain sufficient conditions, his patients were terminally ill, and he used an appropriate drug, an opiate, to 'relieve their suffering'. Similarly in R v Moor [1999], Dr Moor was accused of murdering a patient because he admitted to using lethal doses of morphine on this terminally ill man. Dr Moore was also acquitted, despite his avowed sympathies with euthanasia, because he had used an appropriate drug to treat the suffering of his terminally ill patient. The Court came to a different conclusion in the case of R. v Cox [1992] in which a Doctor Cox was convicted of attempted murder when his patient Lillian Boyes died following an injection of potassium chloride. Cox was convicted because his chosen means, potassium chloride, has no analgesic effect and had no therapeutic uses in the circumstances, thereby falling short of one of the sufficient conditions for a, in Professor Tännsjö's terms, 'tolerated' act, since he could not claim that potassium chloride had a 'double effect'. [9] Had Doctor Cox killed Lillian Boyes with an opiate then it is highly unlikely that his action would have resulted in a conviction. In reporting these cases I make no comment on their moral nature I merely use them to indicate that an action can be seen to meet the conditions of the DDE, placing the act in the class of tolerated interventions when there are reasons to suspect they are acts of a different moral kind. A more relevant example of the 'Cynical Doctrine' is the once familiar, and I believe not entirely eradicated, practice of placing a dying person on a morphine infusion with a prescription for an escalating dose of morphine to be continued until they are dead. I believe that some acts of this kind have cynically employed the DDE to disguise actions that fall into the upper left 'forbidden' quadrant of Professor Tännsjö's matrix. And the most troubling of these are cases in which no reference is made to the patient's wishes. On these grounds I am in agreement with Professor Kuhse and others who argue that the DDE allows euthanasia by subterfuge.

However one way in which the cynical uses of the DDE can be exposed is to focus more closely on the objective conditions of the DDE and hence to focus more closely on the objective aspects of an action as a reliable indicator of the actor's intention. This in turn raises questions about the standards of care and what constitutes an appropriate intervention. Research into palliative care, especially research that has attempted to delineate and clarify palliative interventions, is very relevant here. At the beginning of this section I referred to the ignorance that existed around pain and its effective treatment, particularly by the use of opiates, which led to either ineffective treatment or premature death. However since that time the groundbreaking work of amongst others Robert Twycross[10] into the pharmacokinetics and clinical uses of opiates has standardised the use of opiates for pain and other symptom relief. [11] The principles of appropriate palliative treatment lie in the twin concepts of adequacy and proportionality. Hence an intervention is justified if the treatment is adequate for the task and is proportionate in its effect to the symptoms the patient is experiencing. Robert Twycross and Sylvia Lack, in their handbook 'Oral Morphine in Advanced Cancer'[12] give many practical examples of the principles of managing a patient's pain with morphine. For example morphine is first indicated when a patient experiences pain after being treated with an optimal regime of non-opioid or weak opioid analgesics. Having established that a patient's pain is in fact responsive to opiates then the aim should be to stabilise the patient on a regime that keeps them comfortable with the minimum of side effects. If a patient experiences pain whilst taking morphine then they give the following advice: 'The aim is to increase the dose progressively until the patient's pain is relieved (dose titration). The patient should be advised to increase the dose by 50% if the first dose is not more effective than the previous medication...' (p.11). Through this method it is possible to manage a person, pain free and conscious on a dosage of morphine which would literally be fatal to a person naïve to the drug, indicating that even a large dosage of morphine can be proportionate. This stands in clear contrast to the scenario I described earlier where patients are placed on an escalating dosage of morphine, or the legal cases in which patient's were administered a disproportionate dose of opiate, that is, without reference to the patient's levels of comfort or tolerance of the drug. It is therefore possible to obtain objective evidence of the intention behind an intervention, which may expose important differences between ostensibly similar acts. In Law this possibility has informed the legal standard in negligence, and in the English courts it is known as the 'Bolam Standard' which states that: 'A doctor is not guilty of negligence if he has acted in accordance with a practice accepted as proper by a responsible body of medical men skilled in that particular art.'[13] Of course this leaves open the possibility that there may be differences of opinion between different bodies but in the context of palliative care a great deal of effort has been made to identify and clarify the principles and standards of practice. Whilst this does not render disagreement impossible it shows that it is possible to give evidence of demonstrable consensus, which in the context of pain control makes it possible to expose the cynical exploitation of the DDE.

I have spent some time discussing the principles of palliative care in the context of pain relief because I now wish to go on to show how these principles can be applied to sedation in palliative care. There is however an important caveat since there is nothing like as strong a consensus on what constitutes the 'gold standard' of treatment with regard to sedation, in the way that morphine is to pain.[14] However I believe that the principles of palliative care are robust enough to establish some of the parameters.

There is no doubt that sedation is appropriate in the palliative care context, whether this is to aid in bringing sleep, relieve anxiety or to augment the effects of analgesics. However such uses of sedation fall within the range of normal clinical indications, but *palliative* sedation is employed in circumstances where a reduction in consciousness is the only means of relieving a person's suffering because their symptoms are refractory to the standard treatment. This is consistent with the goals of palliative care, which aims at the patient's comfort rather than cure.

Having discussed the principles of palliative care in relation to pain management it should now begin to be clear why Professor Broeckaert's suggestion that the use of sedation in the palliative care context should be referred to as *palliative* sedation. This is not because the term is specific to all circumstances but rather because it implies specific criteria that must be met in order to justify its use, the key principles of adequacy and proportionality.

Professor Broeckaert offers the following definition: Palliative sedation is 'the intentional administration of sedatives in such dosages and combinations as required to reduce the terminal patient's consciousness as much as needed to adequately control one or more refractory symptoms.' [15] The difficulty I have with the term and the definition of terminal sedation as set out in the initial chapters of this volume is that it describes only the most extreme end of the spectrum of palliative sedation and does not consider the issue of withholding hydration and nutrition as a distinct question. Professor Broeckaert's definition leaves open the possibility that the sedation used might be light or heavy, constant, periodic or temporary. I shall say something further about sedation first before considering the issue of hydration and nutrition.

Any person who is suffering can be relieved of their suffering by being rendered deeply and permanently unconscious. Adopting this blunderbuss approach in all cases will not be sensitive to how people wish to live out their last days – whether an intervention is adequate and proportionate can be judged by attending to a number of issues. First how was the treatment instigated? Were all reasonable measures taken to treat the symptom directly with conventional therapies? Is the decision to sedate doctor or patient initiated? If it was doctor initiated was this with the patient's consent, did the doctor seek appropriate advice from the team or other expertise? If it was patient initiated can their needs be met in alternative ways to their satisfaction?

There are many refractory symptoms, including pain, agitation and dyspnoea, for which sedation may be considered an appropriate palliative intervention. I do not intend to discuss such symptoms here however I do wish to raise the question of whether psychological or *existential* suffering can be legitimately regarded as a symptom for which sedation is indicated. It should be clear from the preceding discussion that I believe this to be the case in the context of terminal illness. It is not difficult to imagine the circumstances in which a person may come to experience the mere awareness of their continued existence as a cause of suffering. As with any other symptom the person's 'self report' must be taken at face value and responded to. Nor is it difficult to envisage that such suffering may also be refractory to a range of interventions of a pharmacological, psychological even spiritual kind. It is in such circumstances that the possibility of using deep and prolonged sedation of the kind envisaged in Professor Tännsjö's definition of terminal sedation. Of course such interventions are not without precedent within the western palliative care context and as such are compatible with and acceptable to the nursing approach to care of the dying. As I have reported, to sedate a person even to the point of unconsciousness is not to abandon the person, since they will be cared for with respect and gentleness in just the same way that a dying person unconscious from natural causes would be cared for.

It might be objected that responding in such a way to existential suffering is characteristic only of some aspects of western medicine and the extremes to which it resorts out of obsessive devotion to patient autonomy. However this objection cannot be sustained. Such practices are also acceptable to cultures normally seen as antithetical to the autonomy-centric Northern European and Anglo-American traditions. Juan Manuel Nunez Olarte, a palliative care physician, and Diego Gracia, a bioethicist, comment that Spain is a country with a relatively mature palliative care service but is also culturally Southern European and Catholic. A feature of this culture is its rejection of euthanasia but acceptance of 'pain relief to the point of sedation, even in cases where death might be accelerated. Within the Catholic tradition, the Thomistic 'principle of double effect', broadly developed by the Spanish 'School of Salamanca' in the sixteenth century is commonly used by Spanish physicians to support the use of analgesics and sedatives.' [16] These authors report that a particular feature of terminal care in Spain is the phenomenon of 'Spanish Death'. This is marked by a readiness to respond to psychological or spiritual suffering with sedation, which '...implies that unconsciousness, either disease-induced or drug induced, is generally perceived as the 'best way out', especially when patients are aware of their prognosis, regardless of whether life is shortened by the use of these drugs' (58).

My conclusion therefore is that sedation even deep sedation, where it is adequate and proportionate, is an appropriate palliative intervention and therefore compatible with nursing values regarding palliative care. Sedation can be initiated in response to refractory physical symptoms where the depth and duration of the sedation is proportionate to the symptoms. Sedation can be doctor instigated with the consent of a competent patient or in the patient's best interests where the patient lacks capacity. Sedation may also be instigated at the patient's request for reasons of psychological or existential suffering, although from a nursing perspective deep continuous sedation of a person who is not imminently dying would be much harder to accept in the hope that less drastic measures, perhaps including light sedation, might be found to make the person's final days comfortable.

6. NUTRITION AND HYDRATION IN THE SEDATED PATIENT

The question of whether to feed or hydrate a sedated patient ought to be regarded as a separate question to that of whether to sedate or not. The model for how to proceed ought to be premised upon that of the patient whose reduced consciousness is disease induced. Many people who are seriously ill and in the terminal phase of a disease have reduced appetite, consuming far fewer calories, and drinking much less liquid than would be considered optimum say in a recovering patient. Managing a patient's nutritional and hydration needs is an integral part of nursing care, but nursing care of the person who is terminally ill is concerned more with responding to his or her desires, in the sense that the primary goal of care is the patient's comfort rather than their biological needs so to speak. [17] What would the person like to eat? Is their mouth dry and if so would sipping some fruit juice or sucking on ice help? Rarely in this context would it be appropriate to attempt to maintain the optimum calorie and fluid intake by artificial means. Because the emphasis in terminal care is on comfort and therefore artificial nutrition and hydration might rightly be regarded as futile because the purpose for which they are indicated is not relevant and they are likely to detract from rather than add to the main goal of terminal care. Of course it is true that no person can survive for very long without food or water but the purpose of terminal care is not to prolong life, and artificially feeding and hydrating in circumstances where a person is no longer able to or has decided not to eat or drink, is likely to do just that. Other things being equal the goal in caring for the imminently dying patient is to achieve something as close to our intuitions regarding a natural death as is possible. Here I agree with Professor Tännsjö that if the patient died sooner than they otherwise would as a consequence of not feeding or hydrating them, then this is indeed an instance of nature taking its course. There is nothing cynical or slight of hand here; from a nursing point of view this is not to abandon the patient. All nursing care would continue, the patient's mouth would be cleaned and moistened at regular intervals, they would be turned and kept clean and spoken to with gentleness and respect.

It is a different matter if we consider the sedation of a patient who was managing to consume some food and water by mouth. If the patient is competent prior to the decision to use sedation then the question of whether to treat with artificial hydration once sedation has occurred ought to be dealt with as a separate decision. This leaves open the possibility that artificial hydration may be utilised alongside sedation. Of course a competent patient is at liberty to refuse a treatment such as artificial hydration, and where this is their competent and informed decision prior to sedation then this is also a situation compatible with nursing care at the end of life.

Greatest difficulty arises in the case of a terminally ill person who requests deep sedation because of their existential suffering, and who has been eating and drinking normally. This scenario is perhaps closest to the extreme case of terminal sedation set out in chapter 2. However even this extreme case may be compatible both morally and legally with nursing care at the end of life.

A person who is terminally ill is not necessarily a person who is imminently dying; the definition of terminal illness allows that a person may have up to a year to live. [18] So what is compatible with good nursing care of a person who is terminally ill but *not* imminently dying? Assuming also that the person is conscious and competent then it is difficult to imagine a set of circumstances in which an ethically or legally justified decision to withhold food and fluids could be made by a doctor or team of health professionals without first obtaining the patient's consent. However to respect a decision made by a competent patient not too eat or drink, or to remove a feeding tube and intravenous catheter if requested to do so would be regarded as ethical and also legally required. [19] This is not to say that most nurses and doctors would not be unsettled by such a decision and would attempt to explore the patient's reasons for making such a request, explore possible alternatives and even perhaps try and dissuade them from this course of action if there were alternatives as yet untried.

On the matter of the patient's request for deep sedation then much will hinge upon whether existential suffering is accepted as a symptom, and whether this symptom is refractory to conventional treatment. Even amongst those who are not sceptical with regard to the validity of existential suffering it would be difficult to find a consensus as to what constituted the standard treatment against which the symptoms can be said to be refractory, anxiolytics, psycho-therapy or spiritual counselling? However on the principle that it is the person him or her self who is the authority as to their own suffering then their own report should be believed and acted upon. Therefore a person who is terminally ill but who is not imminently dying yet complains of existential suffering must be in a position to receive appropriate, that is adequate treatment which is proportionate to their needs; and therefore treatment with deep sedation as a last resort must remain an option. If sedation is combined with a competent refusal of artificial hydration and nutrition – then this seems to present a potential scenario in which terminal sedation of the kind described in chapter 2 may be instigated in a palliative care setting. Of course individual nurses may find this scenario difficult to reconcile with their view of the best that good palliative care can offer. However nursing is not committed to the view that dying people are under an obligation to experience their own dying process nor is it conceivable that imposing such awareness on a dying person, by refusing their request for sedation, could ever be justified.

By now it should be clear that much of what I have said in examining the question of terminal sedation from a nursing perspective is in agreement with Professor Tännsjö's position. However I would be more cautious in the way in which terminal sedation is defined. At best professor Tännsjö's definition describes one extreme end of the spectrum of the justified use of sedation in the terminally ill. For this reason I prefer the term palliative sedation. Where I disagree with Professor Tännsjö's position is over his inclusion of decisions regarding hydration and nutrition as part of the same clinical decision to utilise sedation, this as far as possible should be regarded as a related but distinct decision. I have by no means exhausted all of the circumstances in which decisions to combine sedation with treatment withdrawal may be legitimate. My account could envisage the use of deep sedation combined with the use of artificial hydration but withholding other treatment such as antibiotics and ventilation. Finally, it should not be forgotten that the sight of an infusion of fluids into a patient's arm might carry a symbolic significance way beyond its medical importance. Part of palliative care is also directed toward the care of the family and close ones. Therefore I can imagine circumstances where maintaining such an infusion, where it neither added to nor detracted from the dying person's comfort, is justified simply because it brings comfort to the family. This may be regarded by some as a deceit but it is not clear to me that it is altogether a bad thing.

Simon Woods
Ethics and Lifes Sciences Research Institute (PEALS)
University of Newcastle (UK)

NOTES

[1] For an account of the nursing contribution to palliative care see Woods S (2001) The contribution of nursing to the development of palliative care. In: H ten Have & R Janssens (ed.) *Palliative Care in Europe.* IOS Amsterdam. 133-142.

[2] Saunders C (1967) The care of the dying. *Gerontologica Clinica.* 9. 385, see also Saunders C (1960) Care of the dying. Nursing Times Supplement. Here Saunders not only sets out principles for the care of the dying but argues against euthanasia.

[3] Biswas B (1993) The medicalisation of dying. In D Clark (ed.) *The Future for Palliative Care.* The Open University Press. Buckingham.

[4] See the discussion by Ashby M.A, Stoffell B. (1991) Therapeutic ratio and defined phases of care: proposal of an ethical framework for palliative care. *British Medical Journal.* 302: 1322-4.

[5] Broeckaert B, Nunez-Olarte J.M, (2002a) Sedation in palliative care: facts and concepts. In: ten Have H & D Clark (ed.) *The ethics of palliative care: European perspectives.* Buckingham. Open University Press. 166-180, Broeckart B (2002b) Palliative sedation: ethical aspects. In: C Gastmans (ed.) Between technology and humanity, the impact of technology on health care ethics. Leuven University Press. 239-255.

[6] For a more detailed discussion of these issues see: Woods S, Webb P, Clark D (2001) Palliative care in the United Kingdom. In: H ten Have & R Janssens (ed.) *Palliative Care in Europe.* IOS Amsterdam.85-98, and Clark D, Seymour J (1999) Reflections on Palliative Care. Ch. 5. Open University Press. Buckingham.

[7] For the WHO definition of palliative care see: http://www.who.int/cancer/palliative/definition/en/.

[8] *R v Bodkin Adams* [1957] CLR 365 (CCC) Devlin J

[9] Interestingly Cox did not testify at his trial but Adams did and claimed to be acting only to relieve his patient's suffering.

[10] Twycross R (1990) Therapeutics in Terminal Cancer. Churchill Livingstone. London.

[11] See for example the WHO 'Analgesic Ladder' in Doyle D, Hanks GWC, Macdonald N, (eds) (1994) *The Oxford Textbook of Palliative Medicine.* Oxford University Press. New York.

[12] Twycross R, Lack S (1993) Oral Morphine in Advanced Cancer. (Revised 2nd edition). Beaconsfield Publishers Limited. England.

[13] After the case in which the standard was established: *Bolam v Friern Hospital Management Committee* [1957] 2 All ER 118.

[14] Although the European Association of Palliative Care (EAPC) Research Network is in the process of preparing guidelines on palliative sedation.

[15] Broeckaert B (2000) Palliative sedation defined or why and when terminal sedation is not euthanasia. Abstract 1st Congress RDPC, December 2000, Berlin (Germany) *Journal of Pain and Symptom Management* 20 (6) S58.

[16] Nunez-Olarte J.M, Gracia D (2001) Palliative care in Spain. In: H ten Have & R Janssens (ed.) *Palliative Care in Europe.* IOS Amsterdam. (57).

[17] For example Ciciley Saunders describes terminal care as the point where "comfort rather than curative or even *palliative procedures"* becomes the most appropriate approach. Saunders C (1966) The management of terminal illness. *British Journal of Hospital Medicine.* December: p. 225).

[18] The definition of terminal illness is notoriously difficult see for example Ashby M.A, Stoffell B (1991) Therapeutic ratio and defined phases: proposal of ethical framework for palliative care. *British Medical Journal.* 302. 1322-4.

[19] A refusal to be fed or hydrated artificially is more straightforward in Law since this would be a case of a competent refusal of treatment, c.f. *Re T (Adult: Refusal of Medical Treatment)* [1992] 4 All ER 64 *Re C (Adult: Refusal of Medical Treatment* [1994] 1 All ER 819, *Ms B v An NHS Trust.*

CHAPTER 6

HELGA KUHSE

WHY TERMINAL SEDATION IS NO SOLUTION TO THE VOLUNTARY EUTHANASIA DEBATE

1. ABSTRACT

In an interesting and provocative article, Professor Tännsjö has suggested that terminal sedation at the patient's request might function as an alternative to voluntary euthanasia and resolve the deadlock between those who support voluntary euthanasia, and those who oppose it on the grounds that it infringes the Sanctity-of-Life Doctrine[1]. I disagree.

While terminal sedation would be an acceptable alternative to some supporters of voluntary euthanasia, it would be opposed by traditional supporters of the Sanctity-of-Life Doctrine. As I shall attempt to show, terminal sedation involves the intentional termination of life or killing and would thus constitute euthanasia.

Substantially, I shall argue that the Sanctity-of-Life Doctrine, in its reliance on the Doctrine of Double Effect, is not an appropriate basis for clinical decision-making and the law. In the absence of convincing reasons to the contrary, doctors should be allowed to provide both terminal sedation and euthanasia to terminally or incurably ill patients, at the patients' request.

2. INTRODUCTION

For nine months, between July 1996 and March 1997, doctors in the Northern Territory of Australia were permitted, by law, to practise voluntary euthanasia under stringent conditions laid down in the Rights of the Terminally Ill Act.[2]

When the Act was invalidated by Federal Legislation[3] one terminally ill patient, Ester Wild, had fulfilled all the requirements laid down in the Act, but her doctor, Philip Nitschke, was no longer able to end her life by administering a fast-acting lethal drug. Instead, and with the patient's consent, he commenced a regime of

Torbjörn Tännsjö (ed.), Terminal Sedation: Euthanasia in Disguise?, 57-70.

terminal sedation and Ester Wild died four days later, in a state of medically induced unconsciousness.[4]

Some three out of four Australians support voluntary euthanasia, but with the exception of the short-lived Northern Territory legislation all attempts to legalise the practice have so far failed. In Australia, as in many other parts of the world, a deadlock seems to have been reached in the euthanasia debate. The arguments of those who see voluntary euthanasia as a humane and morally proper choice at the end of life are countered by those who object to it on the grounds that it infringes the inviolability, or sanctity, of all innocent human life. Similar circumstances prevail in many, if not most, Western countries. In light of this, the idea – put forward by Professor Tännsjö – that terminal sedation, at the patient's request, might function as a compromise position and a way out of the deadlock over voluntary euthanasia is an attractive one.

But would terminal sedation be an acceptable alternative to those who find themselves on opposite sides of the voluntary euthanasia debate? Before this question can be answered, it is necessary to take a closer look at the ethical presuppositions underpinning, on the one hand, the Sanctity-of-Life Doctrine and, on the other, the pro-euthanasia position. Only once these presuppositions have been made explicit will it be possible to determine whether terminal sedation can fulfil the role Professor Tännsjö has in mind for it.

For the purposes of this chapter, I shall, following Professor Tännsjö, use the following definitions. 'Euthanasia' is understood as the administration of a non-therapeutic lethal drug by the doctor to a terminally or incurably ill patient, with the explicit intention of ending the patient's life; and 'voluntary euthanasia' as the administration of such drugs at the patient's request.[5] I understand 'terminal sedation' as the deliberate induction and maintenance of deep unconsciousness in a terminally or incurably ill patient until death occurs, coupled with the forgoing of medical treatment, and the withholding of hydration and nutrition,[6] and 'voluntary terminal sedation' signifies the provision of terminal sedation, at the competent patient's informed request. I assume that in all these instances, the doctor is acting in what she sees as the patient's best interests.

3. THE SANCTITY-OF-LIFE DOCTRINE

Professor Tännsjö seems to understand the Sanctity-of-Life Doctrine (or what he sometimes calls the Standard View) as merely prohibiting the intentional termination of life, or killing, by *active* means. That is, however, not how the Doctrine has traditionally been understood. The Sanctity-of-Life Doctrine recognizes that one can kill actively or passively and categorically prohibits the intentional termination of life. It does not, contrary to Professor Tännsjö, subscribe to the acts and omissions distinction as such.[7]

The Doctrine has its source in the Judaeo-Christian tradition, which holds that all innocent human life, irrespective of its quality or kind, is equally valuable and inviolable and must never deliberately be taken. Life is not our own, to do with as we please, but is seen as belonging to God and is 'entirely an ordination, a loan and a stewardship.[8]

It is from the premise that life does not belong to us but to God that the absolute prohibition of the intentional termination of innocent human life is derived. As the Vatican's 1980 *Declaration on Euthanasia* states, 'euthanasia' or 'mercy-killing' is

> an action or an omission which of itself or by intention causes death.[9]

Many people today no longer base their moral thinking on traditional religious precepts; but the attitudes to which the Sanctity-of-Life Doctrine has given rise have become part of common moral thinking, and have been incorporated into professional codes of conduct, public policies and laws.

In line with Anglo-American law, Australian Criminal Codes, for example, recognise that a person can kill, or practise euthanasia, not only by action, but also by omission. As the Criminal Code of Western Australian states:

> A person who does any act or makes any omission which hastens the death of another person who, when the act is done or the omission is made, is labouring under some disorder or disease arising from another cause, is deemed to have killed that person.[10]

On the face of it, these understandings of the sanctity or inviolability of human life would rule out not only euthanasia, but other widely accepted medical end-of-life decisions, such as the forgoing of life-sustaining treatment and the administration of potentially life-shortening pain and symptom control, as well. Terminal sedation involves two medical end-of-life decisions: the induction and maintenance of unconsciousness until death occurs, and the forgoing of all life-sustaining treatments. It will therefore hardly come as a surprise to note that terminal sedation is sometimes straightforwardly regarded as euthanasia or killing in disguise.

Matters are, however, more complex than that. Supporters of the Sanctity-of-Life Doctrine do not advocate the prolongation of life at any cost. Rather, while they reject euthanasia, they take the view that it is permissible to withhold or withdraw so-called 'extraordinary' or 'disproportionate' life-sustaining treatments,[11] and that the administration of life-shortening palliative treatment need not be an infringement of the inviolability of human life. Here the assumption is that these decisions do not, or need not, involve killing or the intentional termination of life.[12]

Similarly the law. In Australia, as in many other countries, competent patients have a recognized common law right to refuse medical treatment for themselves, and doctors who withhold such treatments, are deemed not to have killed their patients, but to have respected their autonomy to be allowed to die. The same is true when it comes to the provision of potentially life-shortening pain and symptom control. It is seen as permissible end-of-life care, rather than as euthanasia.

Now, it may well be the case, as Professor Tännsjö notes, that contemporary mainstream medeical and legal thinking has moved towards the view that the intentional termination of a patient's life by so-called 'passive' means is sometimes ethically accepotable, but this is not the same as showing that such contemporary thinking is in accordance with sthe Sanctity-lf-Life view. Rather, to the extent that medical end-of-life decisions are deemed to involve the intentional termination of life, they would be regarded as impermissible by supporters of the Sanctity-of-Life View — even if they have medical support and are tolerated by the law.

This brings us back to the central question before us: Can the Sanctity-of-Life View consistently reject euthanasia but accept — as Professor Tännsjö has suggested — terminal sedation? I shall argue that the answer is 'no'.

4. EUTHANASIA, FORGOING LIFE-SUSTAINING TREATMENT AND TERMINAL SEDATION

In this Section, I want to focus on one of the two medical-end-of live decisions involved in terminal sedation, the forgoing of life-sustaining treatment, and examine whether and if so how it can be distinguished from euthanasia.

It is widely agreed that euthanasia always involves killing, whereas non-treatment decisions are widely thought of as cases of allowing to die. This distinction between killing and allowing to die is generally believed to rest on the distinction between actions and omissions. But, as Professor Tännsjö correctly notes (pp. 0-0), there is no perfect fit. Killing may not always involve an action, and some omissions may be killings. Take the forgoing of treatment. If the distinction between euthanasia and non-treatment were simply to rest on the acts and omissions distinction, then the withdrawal of treatment (an action) would mean that a doctor who removes a patient from a respirator kills the patient and practises euthanasia, whereas a doctor who does not put the patient on a respirator in the first place merely allows him to die. But this is not how the distinction between impermissible killings and permissible non-treatment decisions has traditionally been understood. Rather, both active and passive end-of-life decisions are sometimes deemed permissible by supporters of the Sanctity-of-Life Doctrine.

Doubts can also be raised with regard to attempts to understand the difference between euthanasia and the forgoing of life-sustaining treatment in terms of the distinction between causing and not causing death. On this view, a doctor who administers a lethal injection is assumed to be causing death and killing the patient, whereas a doctor who discontinues life-support is assumed to merely allow the patient to die. Again, this view is problematical. It rests on an over-simplistic view of causation. As many supporters of the Sanctity-of-Life Doctrine recognize, agents can cause death by either killing or allowing to die. If this were not the case, how could it be that a patient dies when, say, a respirator is turned off, but (other things being equal) stays alive when life-support is continued?

In light of difficulties such as these, the most promising way of attempting to distinguish between euthanasia and forgoing life-sustaining treatment may well be in terms of the difference between initiating a course of events that leads to death, and not intervening in a course of events that, without the intervention, also leads to death.[13] If I understand him correctly, Professor Tännsjö bases his argument in defence of the conceptual difference between euthanasia, on the one hand, and non-treatment decisions, on the other, on broadly this kind of distinction (p. 0). According to this position, euthanasia by way of the administration of a lethal drug would be a case of killing, whereas a non-treatment decision that leads to death would be a case of allowing to die: in the first case the injection initiates a course of events that leads to death; in the second case, the patient will die from an already existing underlying disease, which is not of the doctor's making, and – as it is sometimes put – 'merely allows nature to take its course'.

Let us now apply this understanding of the euthanasia/non treatment distinction to terminal sedation. According to Professor Tännsjö, 'not to feed a patient who, as a consequence, starves to death', is to allow the patient to die (p. 0). This claim is highly problematical. But *even if* one were to grant that not feeding and hydrating a patient can sometimes be regarded as a case of allowing to die, this is not the case when nutrition and hydration are withheld from a patient who has been terminally sedated. In such cases, doctors are not merely standing by 'allowing nature to take its course', rather they have, by first inducing permanent unconsciousness in the patient and then withholding the necessities of life, initiated a course of events that leads to death. On the view we are discussing, they would have killed the patient, not merely allowed her to die.

To illustrate the point: Imagine that Frieda has been given a kitten for Christmas. She puts it in box and deliberately leaves it there, without food and water, until it dies. I think it would be a mistake to say that Frieda has merely allowed the kitten to die. She has killed the kitten: by putting the kitten in the box and leaving it there until it is dead, she has not merely allowed nature to take its course, but has *initiated* a course of events that will lead to death. The same is true in the case of terminal sedation. To put a patient in a state of permanent unconsciousness, without providing hydration and nutrition, is to initiate a course of events that will lead to death and must hence be classified as a case of killing, rather than as a case of allowing to die.

If this is correct, then terminal sedation cannot be accepted by those who hold the view that killing – understood as the initiation of a course of events that leads to death – is always impermissible.

5. SANCTITY OF LIFE AND DOUBLE EFFECT

The belief that it is always impermissible to initiate a course of events that leads to death, that is, engage in a killing, is not uncommon, but it is not the view taken by supporters of the Sanctity-of-Life Doctrine. Rather, those who approach ethics from within that moral framework draw the distinction between impermissible killings and permissible end-of-life decisions in terms of the agent's intentions. If the Sanctity-of-Life Doctrine were to reject as impermissible all actions that amount to the initiation of a course of events that lead to death, then it would have to reject not only euthanasia but also the administration of pain and symptom control that will foreseeably hasten death. This is, however not the case. Rather, as already suggested above, under certain circumstances even the provision of life-shortening palliation is deemed permissible by supporters of the Sanctity-of-Life Doctrine. By the same token, not all omissions are seen as unproblematical. If the agent refrains from acting with the intention of hastening death, then such omissions are regarded as impermissible and, other things being equal, as cases of euthanasia.

In short, the Sanctity-of-Life Doctrine prohibits the intentional termination of life, while at the same time allowing that agents sometimes act, and omit to act, to act in ways that will foreseeably lead to a patient's death.

But when does an agent terminate life intentionally? To answer this question, supporters of the Sanctity-of-Life Doctrine have, as Professor Tännsjö notes (pp. 0-0), traditionally turned to the Doctrine of Double Effect.[14] As the term 'double effect' suggests, one action – such as the administration of pain and symptom control – can have more than one effect: it may alleviate pain *and* hasten death. Provided the death is merely foreseen and not directly intended, and there is proportionality between the bad effect (death) and the good effect (pain relief, or the foregoing of 'extraordinary' or excessively burdensome treatment), the action or omission is permissible.

There are great difficulties clearly explicating the Doctrine of Double Effect and in drawing a clear and consistent distinction between directly intended and merely foreseen consequences. A central problem is that it is often far from clear where the distinction between 'the action' and 'the consequences' should fall.

An example of Jonathan Glover's will illustrate the point:

> When we are on a desert journey and I knowingly use all the drinking water for washing my shirts, my act may be described as one of 'washing shirts', or 'keeping up standards even in the desert', and our being out of water may be thought of as a [merely] foreseen [but not directly intended] consequence.
>
> But it is at least equally acceptable to include the consequence in the description of the act, which may then be described as one of 'using up the last of the water' or of 'putting our lives at risk'.[15]

The same ambiguity prevails when it comes to medical end-of-life decisions. When a doctor administers terminal sedation, for example, the act of sedating the patient may be described as one of 'providing pain relief', with the patient's foreseen death being thought of as a merely foreseen but not directly intended consequence, or as 'killing the patient'.

Similarly in the case of withholding treatment. Turning off a respirator may be described as 'ceasing treatment', with the patient's death being regarded as a foreseen but not directly intended consequence, or as 'bringing about the patient's death'.

How are we to decide what is 'the action' and what are 'the consequences'? Various answers are possible. The Doctrine of Double Effects attempts to provide an answer in terms of the agent's intention. An action, such as the administration of pain and symptom control that is foreseen to bring about death is, in principle, permissible if the doctor's aim is to relieve pain, rather than hasten death. In this case the action would fall under the permissible category 'pain relief', rather than 'killing'; if the doctor were to administer the very same palliative procedure, but with the aim of hastening death, then the action would fall under the description 'intentional killing' and be impermissible.

While I shall later question the wisdom of making the legal permissibility of medical end-of-life decisions depend on the Doctrine of Double Effect, let us see what a plausible interpretation of the Doctrine yields when applied to the practice of terminal sedation.

6. TERMINAL SEDATION

In conjunction with the Doctrine of Double Effect, the Sanctity-of-Life Doctrine will allow the induction of a state of unconsciousness in patients, if that is the only way in which a terminally or incurably ill patient's pain and physical suffering can be controlled, but it would not, I believe, allow the radical withholding of all life-support. Proponents of the Sanctity-of-Life Doctrine have always assumed that only so-called 'extraordinary' or excessively burdensome treatments may be withheld or withdrawn. 'Ordinary' or 'non-burdensome' treatments (exemplified by the provision of food and water), on the other hand, must be provided. If they are omitted, the traditional assumption is that the agent must be acting with the intention of hastening death.[16]

The counter-factual test, often advocated to determine whether agents act with the appropriate intention, yields the same result. According to the test, the agent is asked: 'If you could have achieved the good effect (the relief of pain), without the bad effect (the patient's death), would you still have acted in the way you did?'. If the agent answers the question with 'yes', then the assumption is that she must have acted with the intention of hastening death. A doctor who withholds hydration and nutrition from a patient in a permanent, medically-induced state of unconsciousness would, it seems, have to give an affirmative answer: She could have relieved the patient's pain by the induction of unconsciousness only and would deemed to have acted with the intention ending the patient's life.

That we cannot be sure whether a particular terminally-sedated patient has died as a consequence of heavy sedation, the non-treatment decision, or of 'natural causes' is immaterial here. The Sanctity-of-Life Doctrine and the Doctrine of Double Effect are not concerned with establishing causation, but primarily with the intentions of agents, and the intention of an agent who renders a patient unconscious and then withholds nutrition and hydration must, I believe, be regarded as amounting to the intentional termination of life, or euthanasia.

7. TERMINAL SEDATION: NO SOLUTION TO THE EUTHANASIA DEBATE

If this is correct, then terminal sedation at request will be unacceptable to supporters of the Sanctity-of-Life Doctrine. As a case of the intentional termination of life, or killing, it would, other things being equal, be regarded as the moral equivalent of voluntary euthanasia.

Supporters of voluntary euthanasia, on the other hand, would not generally have any categorical objections to terminal sedation. While it might not make much sense to some members of this group that patients should have to undergo a prolonged period of medically induced unconsciousness and non-treatment prior to death, terminal sedation would, I believe, generally be welcomed as yet another desirable option at the end of life.

This does not mean, however, that all supporters of voluntary euthanasia will regard terminal sedation as a satisfactory alternative to voluntary euthanasia. Terminal sedation would satisfy those who base their arguments for voluntary euthanasia on an experience-based model of welfare (where the avoidance of pain and suffering at the end-of-life would be central).[17] It would, however, be unlikely to satisfy those who take the view that patients, like moral agents in general, have not only experiential interests but also an interest in controlling and shaping their lives in accordance with their preferences or self-chosen life-plans and goals.[18] A good death – euthanasia – some might argue, is not so much a pain-free death, as a death that reflects the patient's deeply held values and beliefs.[19]

This analysis strongly suggests that terminal sedation at request cannot fulfil the role Professor Tännsjö somewhat optimistically envisages for it. Rather than accept it as a reasonable compromise position, supporters of the Sanctity-of-Life Doctrine are likely to reject it as a form of euthanasia; proponents of voluntary euthanasia, on the other hand, are likely to regard it as desirable additional option at the end of life, but not all would accept it as a true alternative to voluntary euthanasia.

Professor Tännsjö's argument for the compatibility of voluntary terminal sedation with the Sanctity-of-Life Doctrine is clearly not intended as a merely abstract philosophical exercise; rather, it has the practical intent of ensuring that patients need not die in 'unnecessary agony, pain and distress'(p. 0). If I am correct, Professor Tännsjö's argument fails. Supporters of the Sanctity-of-Life view will reject terminal sedation and many incurably and terminally ill patients will continue to be unable to determine when and how they die.

8. REJECTING THE SANCTITY-OF-LIFE DOCTRINE

In arguing for his conclusion, Professor Tännsjö did not question the Sanctity-of-Life Doctrine; he accepted it as given. But the Sanctity-of-Life Doctrine needs to be questioned. In meeting Professor Tännsjö on his own ground, my chapter has so far focused on consistency, that is, on whether terminal sedation is compatible with the conceptual and ethical presuppositions that underpin the Sanctity-of-Life Doctrine and the pro-voluntary euthanasia position. I have not discussed substantive questions of ethics, and ignored issues surrounding public policies and laws. There are, I have argued elsewhere, good philosophical and ethical grounds for rejecting the Sanctity-of-Life Doctrine,[20] and for permitting, as a matter of public policy and law, the practice of voluntary euthanasia.[21] For the purposes of this chapter, a brief sketch of some of the arguments for the rejection of the Doctrine as a basis for clinical practice and public policy must suffice.

The Sanctity-of-Life Doctrine has its source in the religious belief that the intentional termination of human life is always impermissible, but that it is sometimes permissible to allow a human being to die. Supporters of the Doctrine see the distinction between what an agent directly intends and what she merely foresees as having profound moral significance, while others – such as supporters of voluntary euthanasia – often regard it as a hypocritical device that allows those who subscribe to absolute rules to avoid the starkly counter-intuitive consequence of a strict adherence to those rules. After seemingly endless debates, no consensus is in sight and none seems likely. The reason is that questions relating to moral significance can often not be answered other than from within particular moral perspectives. Different moral perspectives will give rise to different answers – and these answers cannot be shown to be true or false, in the ordinary sense of those terms.[22]

While the question of the *intrinsic* moral significance of the intention/foresight may thus be ultimately be unanswerable, there appear to be good *extrinsic* reasons why the distinction should - in the absence of overriding reasons to the contrary – be rejected as a basis for clinical practice and public policies and laws that regulate medical-end-of-life decisions. Here I want to mention but five.

i. *The problem with intention: clinical practice*. At the beginning of this article, I mentioned the case of Ester Wild, who died after having undergone terminal sedation. Following Ester Wild's death, her doctor, Philip Nitschke, was subject to a coronial investigation.[23] The question was: did he or did he not act with the intention of shortening his patient's life when he provided terminal sedation?

We cannot be sure what Dr. Nitschke's intentions were, or what other doctors intend, or aim at, when they implement medical end-of-life decisions. Intentions are subjective mental states, and often only the agent her- or himself will be able to say what her intentions were. Nor will appeal to the Doctrine of Double Effect always be conclusive. The Doctrine is a complex set of principles, and can give rise to different interpretations.[24]

As a consequence, writers in the field have argued, reliance on the Doctrine of Double Effect creates serious problems for clinical practice.[25] Roger Hunt, a medical practitioner who has pioneered hospice and palliative care in South Australia points out that the Doctrine of Double Effect is not always able to help doctors to distinguish between 'euthanasia' and 'palliative care' in a consistent and practically useful way. The Doctrine of Double Effect rests on an implausibly narrow notion of intent, and fails to recognise the philosophical and psychological complexity of clinical intentions. Those involved in the care of dying patients can, and often do, act with a range of different intentions. Palliative care and euthanasia are not, he argues, distinct practices; rather they lie on a continuum of end-of-life decisions, where some forms of palliative care might be described as cases of 'slow active euthanasia'.[26]

While there is thus not, Hunt continues, a clear dividing line separating permissible end-of-life decisions from impermissible ones, the continuing belief in the existence of such a line stands in the way of good patient care; it will foster and reinforce self-deception, secrecy and isolation, undercut professional responsibility, and may lead to the abandonment of patients when patients are most in need of care.[27]

ii. *Discrimination*. Competent patients have a widely recognized legal right to refuse treatment, including life-sustaining treatment, and doctors are permitted, indeed legally required, to act on their patients' refusals. This gives some patients – those in need of life-support – the opportunity to bring about their own death, with the help of their doctor. Patients who do not need life-support are almost everywhere denied this opportunity.[28] The reason is that the law, like the Sanctity-of-Life Doctrine, prohibits the intentional termination of life. The legal assumption is that doctors who administer a lethal drug at a patient's request are intentionally terminating that patient's life; whereas doctors who discontinue life-support at the patient's request, merely allow the patient to die – even if they clearly foresee, and intend,[29] that the patient will die as a consequence of their action.

In light of this, patients who do not need life-support might want to argue that the continued legal prohibition of voluntary euthanasia is discriminatory and unjust. Existing laws allow one group of patients – those who are fortunate enough to require life-support – to bring about their own deaths, with the help of their doctors. Another group of patients – those who are suffering just as much, but are in the unfortunate position of not requiring life-support – are denied that opportunity. This means, such an argument might continue, that existing laws are not treating patients fairly; they show partiality to one group of dying patients, while disregarding the equal interests or rights of another group, on the basis of the arbitrary criterion that patients in the first group do, and patients in the second group do not, require life-support which they can legally refuse.[30]

iii. *Is there a slippery slope to non-voluntary euthanasia?* In response to such arguments for voluntary euthanasia it is often said that there are good reasons for not allowing the practice. Arguments for the annulment of the *Northern Territory Rights of the Terminally Ill Act*[31] claimed that the acceptance of voluntary euthanasia would lead to non-voluntary euthanasia, where doctors would intentionally be ending the lives not only of those who are terminally or incurably ill and want to die, but also the lives of those who cannot or will not give their consent.

Available empirical evidence points in the opposite direction. Recent sophisticated international studies across three countries have shown that the public acceptance and regulation of voluntary euthanasia may reduce, rather, than increase the incidence of non-voluntary euthanasia. In Australia and Belgium, where voluntary euthanasia is unlawful, the incidence of non-voluntary euthanasia was found to be around five times higher than in the Netherlands, where voluntary euthanasia has been allowed, and regulated, since the early 1980's. In the Netherlands, the incidence of non-voluntary euthanasia was found to be 0.7% of all deaths; in Australia it was 3.5% and in Belgium 3.2%.[32] Quite clearly, these data offer no support for the slippery slope argument against voluntary euthanasia. On the contrary, they suggest that if there is a 'slippery slope', it slopes in the opposite direction: the acceptance and regulation of voluntary euthanasia is linked with a lower incidence of non-voluntary euthanasia.

iv. *The problem with intention: public policy.* As we have seen, under the Doctrine of Double Effect, it is the doctor's intention that determines the permissibility or impermissibility of non-treatment and palliative care decisions that are foreseen to have a life-shortening effect. This reliance on doctors' intentions makes for what has been termed the 'malleability' and 'constructability' of medical end-of-life decisions.[33] Doctors who do not want to bring about a patient's death by administering a non-therapeutic drug, will often have other means of achieving the same result: they may intentionally bring about death by increasing pain and symptom control, and/or by way of non-treatment decisions. According to the above surveys, a total of 36.5% of all Australian deaths, 19.5% of Belgian deaths and 19.5% of Dutch deaths (1990 survey) were preceded by a medical end-of-life decision that was explicitly or partially intended to shorten the patient's life; but only 5.4% (Australia), 4.5% (Belgium) and 3.3% Netherlands involved euthanasia; the remainder involved non-treatment decisions and the administration of drugs used in pain and symptom control.[34]

These findings make clear that laws prohibiting doctors from intentionally shortening the lives of some of their patients are not being observed. While prohibitive laws may prevent some doctors from using fast-acting non-therapeutic drugs, they do not, and cannot, prevent doctors from intentionally ending their patients' lives by some other means and may, in fact, encourage – rather than discourage – the unconsented-to termination of life. Indeed, one might reasonably predict htat if terminalo sedation were accepted as a morally and legally permissible option, against the backdrop of the continued prohibition of the intentional termination of life, there would be an increase — not decrease — in the *unconsented-to* termination of life.

For all the reasons outlined in this Section, and in the absence of convincing reasons to the contrary, a change in focus is sorely needed. When medical end-of-life decisions for competent patients are at issue, we should stop asking whether a doctor 'intends' death or merely 'allows' it to occur, whether death comes about as the result of an action or an omission, or as the result of doctors administering a slow-acting therapeutic or a quick-acting non-therapeutic drug. While some of these distinctions have moral relevance in the context of some religious or moral views, they are not a proper basis for clinical practice and public policies regulating medical end-of-life decisions. What is needed is a single regulatory framework for all medical end-of-life decisions for competent patients, a framework that does not rely on the largely unworkable notion of intention, but on the substantive notion of respect for the patient's autonomy, which finds expression in the procedural requirement of consent. Under such a framework, patients and doctors would be free to jointly decide on a mode of dying that best meets the needs of the patient. For most patients this would involve forgoing life-sustaining treatment and recourse to standard palliative care. But for some patients, this would also involve terminal sedation and voluntary euthanasia.

9. CONCLUSION

Against Professor Tännsjö I have argued that terminal sedation is not compatible with the precepts on which the Sanctity-of-Life Doctrine is based. From the vantage point of that Doctrine, terminal sedation is the moral equivalent of euthanasia and is likely to be rejected by consistent adherents of that doctrine. Most supporters of voluntary euthanasia, on the other hand, would probably welcome terminal sedation as another desirable option at the end of life, but would not regard it as an acceptable alternative to voluntary euthanasia.

Substantively, I have argued that the mainstream legal assumption that doctors who administer fast-acting non-therapeutic drug at a patient's request practise euthanasia, whereas doctors who withhold or withdraw life-sustaining treatment and administer life-shortening palliative care do not, offers neither transparency nor regulatory protection. It does not encourage honesty and openness in the doctor/patient relationship and may encourage doctors to act paternalistically, rather than with their patients' consent.

Based on respect for their autonomy, competent patients have a legal right to refuse life-sustaining for themselves. I believe that in keeping with this tradition, medical end-of-life decisions in clinical practice and law should be based on, and guided by, patients' informed consent, rather than the Doctrines of the Sanctity of Life or Double Effect.

Helga Kuhse
Monash Centre for Human Bioethics
Monash University

NOTES

[1] "Terminal Sedation – A Compromise in the Euthanasia Debate?", *Bulletin of Medical Ethics*, No. 163, November 2000, pp. 13–22. He has developed his argument further in Chapter 2 of this book, and all my subsequent references are to this chapter.
[2] Rights of the Terminally Ill Act 1995, Northern Territory of Australia. Darwin: Government Printer, 1995.
[3] Commonwealth Euthanasia Laws Act 1997.
[4] Terminal sedation is practised in Australia, but the moral and legal status of various regimes described by the term remains contentious. For some backgrouund on the case of Ester Wild and the Northern Territory Legislation, see David W. Kissane, Annette Street, Philip Nitschke: 'Seven deaths in Darwin: case studies under the Rights of the Terminally Ill Act, Northern Territory, Australia', *The Lancet*, Vol. 352, October 3, 1988, pp. 1097-1102. See also Helga Kuhse: 'Voluntary euthanasia, politics and the law', *Monash Bioethics Review*, Vol. 16, No.1, January 1997, pp. 1-4; also 'Death of the NT Rights of the Terminally Ill Act', *Monash Bioethics Review*, Vol. 16, No. 3, July 1997, pp. 3.
[5] Professor Tännsjö does not define voluntary euthanasia in this way, but I take it that this is the understanding he has in mind when, on p. 17, he refers to the Dutch practice of voluntary euthanasia as 'the active and intentional killing of patients ... at the request of these patients ...'
[6] Professor Tännsjö's initial definition of 'terminal sedation' covers only terminally ill patients, but is then extended to cover incurably ill patients as well (p. 26).
[7] For a detailed discussion and critique of the Sanctity-of-Life Doctrine in Medicine, see Helga Kuhse: *The Sanctity-of-Life Doctrine in Medicine - A Critique*, Oxford: Oxford University Press, 1987.
[8] Paul Ramsey, as quoted by Daniel Callahan: 'The Sanctity of Life' in Dr. R. Cutler (ed.): *Updating Life and Death*, Boston: Beacon Press, 1969, pp. 181-223; see also Paul Ramsey: *Ethics at the Edges of Life*, New Haven and London: Yale University Press, 1978, p. 147.
[9] Sacred Congregation for the Doctrine of the Faith: *Declaration on Euthanasia*, Vatican City, 1980, p. 6; in the Appendix of this book, p. 136.
[10] Criminal Code of Western Australia, Section 273. See also the New South Wales (Australia) Crimes Act 1900, S.19(1)(a) which states: 'Murder shall be taken to have been committed where the act of the accused or *the thing omitted by him to be done, causing the death charged,* was done or omitted with reckless indifference to human life, or with intent to kill or inflict grievous bodily harm upon some person.' (my emphasis) For a general discussion of the sanctity of life view in law, see E.W. Keyserlingk: *Sanctity of Life or Quality of Life (in the context of ethics, medicine, and law).* Study written for the Law Reform Commission of Canada, Ottawa: Law Reform Commission, 1979.
[11] Sacred Congregation for the Doctrine of the Faith, op. cit., p. 10; reprinted in the Appendix of this book, p. 138.
[12] Sacred Congregation for the Doctrine of the Faith, op. cit., pp. 8-9; reprinted in the Appendix of this book, p. 139. For a critical analysis of these assumptions, see Helga Kuhse: *The Sanctity-of-Life Doctrine in Medicine - A Critique,* op. cit.
[13] See Chapter 2, Helga Kuhse: *The Sanctity-of-Life Doctrine in Medicine - A Critique,* op. cit.
[14] For a statement of the Doctrine, see Catholic University of America: *New Catholic Encyclopedia*, Vol. 4, New York: McGraw Hill, 1976, pp. 1020-22.
[15] Jonathan Glover: *Causing Death and Saving Lives*, Harmondworth: Penguin Books, 1977, p. 90.
[16] Sacred Congregation: *Declaration on Euthanasia*, op. cit., p. 10; reprinted in the Appendix of this book, p. 137..
[17] Hedonistic utilitarians would fall into this group.
[18] Preference utilitarians and various rights or autonomy-based ethical theories would take this kind of view.
[19] For two such views, see, for example, Max Charlesworth: *Bioethics in a Liberal Society* (Cambridge: Cambridge University Press, 1993); Ronald Dworkin: *Life's Dominion: An argument about abortion, euthanasia and individual freedom*, New York: Knopf, 1993.
[20] *The Sanctity-of-Life Doctrine in Medicine*, op.cit.
[21] See, for example, Helga Kuhse: Chapter 8, Helga Kuhse: *Caring: Nurses, Women and Ethics*, Oxford: Blackwell, 1997; and Helga Kuhse: 'From Intention to Consent' in (eds.) Margaret P. Battin, Rosamond Rhodes, and Anita Silvers: *Physician-Assisted Suicide - Expanding the Debate*, New York and London: Routledge, 1998, pp. 252-266.

[22] See H.T. Engelhardt: *The Foundations of Bioethics*, second edition, Oxford: Oxford University Press, 1995.
[23] The investigation was subsequently dropped.
[24] See Chapter 3, Helga Kuhse: *The Sanctity-of-Life Doctrine in Medicine*, op. cit.
[25] See, for example, Roger Hunt: 'Palliative Care: The Rhetoric-Reality Gap', In Helga Kuhse (ed.): *Willing to Listen - Wanting to Die*, Ringwood, Vic.: Penguin Books, 1994, pp. 115-136; and T.E. Quill, R. Dresser, D.W. Brock: 'The Rule of Double Effect - A Critique of its Role in End-of-Life Decision Making', *New England Journal of Medicine*, Vol. 327(24), 11 December 1997, pp. 1768-1771.
[26] Roger Hunt: 'Palliative Care: The Rhetoric-Reality Gap', In Helga Kuhse (ed.): *Willing to Listen - Wanting to Die*, op.cit., pp. 115-136.
[27] loc.cit.
[28] The Netherlands, Belgium, and Oregon are exceptions. In the US State of Oregon, physician-assisted suicide is lawful, and doctors in the Netherlands and Belgium can lawfully practise both voluntary euthanasia and medically assisted suicide.
[29] P.J. van der Maas, J.J.M. Delden, L. Pijnenborg: 'Euthanasia and Other Decisions Concerning the End of Life', *Lancet*, Vol. 338, 1991, pp. 669-74; P.J. van der Maas, G. van der Wal, I. Haverkate, et al.: 'Euthanasia, Physician-Assisted Suicide, and Other Medical Practices Involving the End of Life in the Netherlands 1990-1995', *New England Journal of Medicine*, Vol. 335, 1996, pp. 1699-1705; Helga Kuhse, Peter Singer, Peter Baume, et al: 'End-of-Life Decisions in Australian Medical Practice', *Medical Journal of Australia*, Vol. 166, 1997, pp. 191-96; Luc Deliens, Freddy Mortier, Johan Bilsen, et al: 'End-of-Life Decisions in Medical Practice in Flanders, Belgium: A Nationwide Survey', *Lancet*, Vol. 356, 2000, pp. 1806-1811.
[30] *Vacco v. Quill*, 80f, 3rd. 716 (2nd Circ) 1966. This case of discrimination was upheld by the Appeals Court, but not subsequently by the United States Supreme Court (*Vacco v. Quill*, 117 S.Ct. 2293, 1997).
[31] See endnote 2.
[32] P.J. van der Maas, J.J.M. Delden, L. Pijnenborg: 'Euthanasia and Other Decisions Concerning the End of Life', *Lancet*, Vol. 338, 1991, pp. 669-74; P.J. van der Maas, G. van der Wal, I. Haverkate, et al.: 'Euthanasia, Physician-Assisted Suicide, and Other Medical Practices Involving the End of Life in the Netherlands 1990-1995', *New England Journal of Medicine*, Vol. 335, 1996, pp. 1699-1705; Helga Kuhse, Peter Singer, Peter Baume, et al: 'End-of-Life Decisions in Australian Medical Practice', *Medical Journal of Australia*, Vol. 166, 1997, pp. 191-96; Luc Deliens, Freddy Mortier, Johan Bilsen, et al: 'End-of-Life Decisions in Medical Practice in Flanders, Belgium: A Nationwide Survey', *Lancet*, Vol. 356, 2000, pp. 1806-1811.
[33] John Griffiths: 'The Regulation of Euthanasia and Related Medical Procedures that Shorten Life in the Netherlands', *Medical Law International*, Vol. 1, 1994, pp. 137-58.
[34] See Helga Kuhse, Peter Singer, Peter Baume, et al: 'End-of-Life Decisions in Australian Medical Practice', op.cit.

CHAPTER 7

DAN W. BROCK

TERMINAL SEDATION FROM THE MORAL RIGHTS' PERSPECTIVE

1. INTRODUCTION

Is terminal sedation morally permissible, and how is it related to forgoing life support, physician assisted suicide and voluntary active euthanasia? I shall follow Torbjorn Tännsjö in Chapter 2 in this volume in understanding terminal sedation as containing two parts: first, sedating the patient to the point of unconsciousness and, second, stopping the provision of nutrition and hydration. Tännsjö argues that terminal sedation is both morally and legally permissible and should be accepted by those who oppose both physician assisted suicide and voluntary active euthanasia. I agree that it is morally and legally permissible. In the United States it is accepted practice in many health care institutions and hospices and so far as I am aware it has not been legally challenged. However, I do not agree with the analysis and argument that Tännsjö gives in support of that position and I want first to delineate our disagreement.

2. TÄNNSJÖ'S APPLICATION OF STANDARD DISTINCTIONS TO END OF LIFE CARE DECISIONS AND ACTIONS

Tännsjö's central argument both for terminal sedation's moral and legal permissibility rests in his appeal both to the active/passive distinction and to the doctrine of double effect which draws a distinction between consequences that are intended versus foreseen but unintended. He argues that in the law (and I think he wants to say morally as well) only killings that are active and intended are legally impermissible and 'inherently wrong.' The other three possibilities — active/unintended, passive/intended, and passive unintended — are sometimes permissible and sometimes impermissible depending on 'their consequences.' There

Torbjörn Tännsjö (ed.), Terminal Sedation: Euthanasia in Disguise?, 71-79.

are two problems with this argument. First, the way he applies these distinctions to different end of life care is non-standard. That would of course be acceptable if their standard application was confused or mistaken. But since he is using these distinctions to interpret standard views on these matters, especially among medical practitioners, in order to show why those who hold standard views should accept terminal sedation, it won't do to interpret them in a non standard and idiosyncratic manner. Second, no argument is given as to why killings that are active and intentional are inherently wrong, whereas killings that have only one or the other of these properties are not. Moreover, I believe and shall argue briefly below that active and intentional killings are not inherently wrong on the most plausible account of the morality of killing.

Let me pursue the first of the two difficulties in Tännsjö's argument. He offers no analysis of the active/passive distinction but suggests nevertheless that its application is clear in the various end of life decisions and actions in question. However, I believe that is not so. Its application is controversial in forgoing life support cases. The active/passive distinction is often taken to be the same as the killing/allowing to die distinction, and many do believe that all forgoing life support is passive or allowing to die. However, in other places I have used the case of the greedy son to call that into question:

> Consider the case of a patient terminally ill with ALS disease. She is completely respirator dependent with no hope of ever being weaned. She is unquestionably competent but finds her condition intolerable and persistently requests to be removed from the respirator and allowed to die. Most people and physicians would agree that the patient's physician should respect the patient's wishes and remove her from the respirator, though this will certainly cause the patient's death. The common understanding is that the physician thereby allows the patient to die. But is that correct?
>
> Suppose the patient has a greedy and hostile son who mistakenly believes that his mother will never decide to stop her life-sustaining treatment…Afraid that his inheritance will be dissipated by a long and expensive hospitalization, he enters his mother's room while she is sedated, extubates her, and she dies. Shortly thereafter the medical staff discovers what he has done and confronts the son. He replies, 'I didn't kill her, I merely allowed her to die. It was her ALS disease that caused her death.' I think this would rightly be dismissed as transparent sophistry — the son went into his mother's room and deliberately killed her. But, of course, the son performed just the same physical actions, did just the same thing, that the physician would have done. If that is so, then doesn't the physician also kill the patient when he extubates her?[1]

There are of course important moral differences in the physician's and the son's actions—differences of consent by the patient, motive, and social role, but these seem not differences in whether each kills or allows to die. So it is not clear and uncontroversial that all forgoing life support is passive or allowing to die, as Tännsjö suggests. Moreover, without an analysis of the difference it is not possible to assess what moral importance, if any, the distinction has, whether alone or in combination with the intended/foreseen distinction.

At least as important is how the intended foreseen distinction applies to these different end of life actions.[2] Tännsjö claims that some cases of forgoing life support are cases of intending death, as are all cases of withholding nutrition and hydration. Here, it is important to distinguish the patient's or surrogate's intention from the physician's intention. The patient may often stop life support, whether nutrition and hydration or other forms, with the intent of ending her life. Likewise, the surrogate's intent in such cases may be ending the patient's life. But does the physician intend the patient's death in such cases. I believe the correct answer is sometimes. But the standard view of physicians and many others is that the physician does not in such cases intend the patient's death. Rather, the physician's intent is to respect the patient's or surrogate's wishes and their right to decide about and to refuse treatment. This applies to life support generally as well as to the second part of terminal sedation which is the stopping of nutrition and hydration. That no withholding or withdrawal of life support by the physician is considered intending the patient's death is reflected in the United States in the American Medical Association's oft stated view that physician's should respect patient's or surrogate's decisions to forgo life support, but that they should never intentionally end patient's lives or intend a patient's death. Moreover, it is widely agreed that sedating the patient to the point of unconsciousness in order to relieve her pain or suffering does not involve intending the patient's death. Even if the sedating should hasten the patient's death, that is a foreseen but unintended consequence of relieving her pain or suffering. The fact that death is not intended in terminal sedation is sufficient to make it in Tännsjö's account morally and legally permissible. Thus, to summarize this section I believe the distinctions to which Tännsjö appeals are not so clear as he suggests, and that standard moral thinking does not apply them to end of life actions in the way that he has done in his paper.

3. WHAT IS THE CORRECT ACCOUNT OF THE MORALITY OF KILLING?

Tännsjö holds that morally and legally only active and intentional taking of life is impermissible. I believe this is reasonably accurate regarding the law in the United States and in many European countries. But is this correct about the morality of killing? In Tännsjö's view, an active taking of life is only sometimes wrong, depending on consequences, and an intentional taking of life is only sometimes wrong, again depending on consequences. But a taking of life that is both active and intentional is 'inherently wrong.' It is not inconsistent to hold that neither condition by itself is sufficient to make an action wrong, but that together the two conditions are sufficient to make it wrong.[3] But an argument is needed as to why these two conditions make an action always wrong, and so far as I can see Tännsjö has not offered one. However that may be, is the view correct? I believe it is not. The fundamental question is what is the correct account of the morality of killing, of taking human life. This is quite clearly an enormously complex issue and I cannot offer anything approaching a full account here, but I can at least indicate in broad outline how my view differs from Tännsjö's.[4] A view that holds that active and intentional taking of human life is inherently wrong is what philosopher's have called a duty-based view.[5] It focuses on the agent and the nature of his action, defining some actions as always, or nearly always, impermissible. A Kantian moral view that requires always respecting other persons as rational agents and treating them as ends in themselves, never solely as means, is a prominent example of a duty-based view.[6] The position of Roman Catholic moral theology that the deliberate killing of innocent human beings is always wrong is another.[8] In both of these views, such actions are wrong even if the victim has consented to being killed and may reasonably want to die. Consent in this view is understood not as authorizing an impermissible action, but as a request or even temptation to do wrong or evil. This is why many Kantians as well as Roman Catholic moral theologians oppose both physician assisted suicide and voluntary active euthanasia.

However, I believe the duty-based view misplaces the focus of moral concern. That concern should be placed first and foremost on the potential victim, as a rights-based view does, not the agent, as a duty-based view does. If we ask why killing persons is typically wrong, I believe the answer must focus on the harm and wrong done to the victim. The harm is the denial to the victim of his future and of all the valued experiences that his future would otherwise have included. The wrong to the victim is taking from him his life – something that is rightfully his, and his to control and dispose of – without his consent. The form of moral view that fits this account of why killing persons is morally wrong will be rights-based.[8] It posits that persons have a moral right not to be killed, but that right, like other rights, is theirs to use as they see fit so long as they are competent to do so; specifically, persons can waive their right not to be killed when they judge that their future is not on balance a good to them, and so its loss will not constitute a harm. In such cases, when they waive their right not to be killed by refusing life support, or by requesting terminal sedation, physician assisted suicide, or voluntary active euthanasia, these actions taken by their physicians (or others) will not wrong them by taking from them without their consent what is rightfully theirs, that is will not violate their right not to be killed. Nor will these actions harm the patient who reasonably judges further life to be on balance a burden and unwanted.

In this rights-based view of the morality of killing, it will be the voluntariness of the patient's request for forgoing life support, terminal sedation, physician assisted suicide, or voluntary active euthanasia that is crucial to their moral permissibility. The rights-based account gives fundamental importance to individual self-determination or autonomy. This view sees patients as having a right to determine the time and manner of their death, a right affirmed as a constitutional right by the United States 9th Federal Circuit Court of Appeals[9], though later rejected by the United States Supreme Court.[10] In this rights-based view of the morality of killing terminal sedation will be morally permissible, but so will physician assisted suicide and voluntary active euthanasia. Whether the killing is active or passive, intended or unintended but foreseen, will not be important to whether it is morally permissible. I cannot defend this alternative rights-based account of the morality of killing here in place of Tännsjö's duty-based account, but I believe it is the more plausible account of the morality of killing.

4. HOW DOES TERMINAL SEDATION FIT STANDARD MORAL DISTINCTIONS?

Perhaps it will be useful to summarize where Tännsjö and I disagree about how terminal sedation fits the usual distinctions employed in morally evaluating end of life care, despite the fact that we both agree that it is morally and legally permissible. Consider the first part of terminal sedation—sedating the patient to the point of unconsciousness, even if this alone may hasten death. Tännsjö and defenders of the doctrine of double effect hold that this is permissible because the risk of earlier death is foreseen but unintended, even if this falls on the active side of the active/passive line. I believe as well that this is permissible, indeed required, treatment if the patient requests it or consents to it, knowing that it carries the risk of hastening her death. The law too is in most countries, and certainly in the United States, quite clear that this form of pain relief is permitted. Now consider the second part of terminal sedation—withholding nutrition and hydration knowing that this will cause the patient's death. Tännsjö holds that this too is permitted because though death is intended, withholding or withdrawing treatment falls on the passive side of the active/passive line. I believe the more common view within the medical profession and among persons generally is that the physician in so acting intends to respect the patient's/surrogate's wishes, foreseeing but not intending that in doing so death will result. In my view, the justification for the physician's action is that it respects the patient's/surrogate's right to decide about and to refuse any treatment, including life-sustaining treatment. The physician's action is in response to the patient/surrogate having waived the patient's right not to be killed. The patient or surrogate, as well as the physician, may or may not intend the patient's death, but that in my view is not morally important to whether the physician's action is morally permissible.

Why have some held that terminal sedation is in essence no different than voluntary active euthanasia, is 'euthanasia in disguise?'[11] Because it looks as if once the patient is sedated to the point of unconsciousness, she can no longer suffer and so the only reason for then withholding or withdrawing nutrition and hydration is to end the patient's life. The patient may well instruct that nutrition and hydration be withdrawn in order to end her life, but so do many conscious patients when they forgo life-sustaining treatment. Stopping eating and drinking is an option of competent, conscious patients, and that too is not euthanasia or assisted suicide in disguise. It is, I believe, principally because many find it hard to see how there cannot be an intent to end the patient's life in terminal sedation when nutrition and hydration is stopped. But I have already argued that neither the patient nor the physician need have this intent in terminal sedation and, moreover, the presence of that intent is not enough to establish that terminal sedation is euthanasia in disguise since that same intent is sometimes present in forgoing life support which few oppose or claim to be euthanasia in disguise. To put the point differently, if terminal sedation is euthanasia in disguise, then so is much forgoing life support.

5. WHAT SHOULD PUBLIC POLICY BE ABOUT TERMINAL SEDATION?

Since terminal sedation is not in itself morally wrong, should public and legal policy permit it? I would add that if, as I believe, physician assisted suicide, voluntary active euthanasia, and voluntary stopping eating and drinking are also not in themselves morally wrong, the same policy issue arises regarding them. It might be thought that the answer to this question is straightforwardly yes — if any of these are morally permissible, they should be legally permissible. But I believe that does not follow. If there is a moral right to control the time and manner of one's death that includes any or all of these practices, that will establish a strong moral presumption that they should not be legally prohibited. But that right and presumption might be overridden by sufficiently bad consequences of permitting them, for example if doing so led to many patients being killed who did not wish to die. Thus, we need to address what the consequences of a public policy that permits these practices would likely be. In particular, many who oppose physician assisted suicide and voluntary active euthanasia grant that they are not in themselves morally wrong, but nevertheless argue that they should not be permitted because of their potential for well-intentioned misuse and ill-intentioned abuse. If terminal sedation should be permitted, it might be because it is less subject to abuse than are physician assisted suicide and voluntary active euthanasia. To answer this question we first need to clarify what would count as abuses. In the account I sketched above of the wrongness of killing, abuses will be actions that take a person's life without his or her informed and voluntary consent. Moreover, I believe it is widely, though of course not universally, accepted that abuses of physician assisted suicide and voluntary active euthanasia are when they take place contrary to a patient's wishes.

If this is correct about what count as abuses, then the dividing line between practices more and less subject to abuse is not where it is often assumed to be in debates about end of life care. The common assumption is that physician assisted suicide and voluntary active euthanasia are more subject to abuse than forgoing life support or pain relief that may hasten death. But the most important division between practices less and more subject to abuse is whether a competent patient is making the decision for him or herself, or whether someone else, such as a surrogate, is making the decision for an incompetent patient. Many studies have shown that other persons such as family members or physicians of patients are not reliable predictors of patients' wishes regarding end of life care, and so reliance on surrogate decision makers instead of competent patients inevitably introduces a significant potential for choices to be made that would be contrary to the patient's wishes.[12] But of course decisions to forgo life-sustaining treatment and to employ pain relief that may hasten death are not only made by competent patients, but are also often made by surrogates for incompetent patients. Decisions to employ either physician assisted suicide or voluntary active euthanasia, on the other hand, should always made by competent patients. That means that in this important respect forgoing life support and pain relief that may hasten death are more, not less, subject to abuse than are physician assisted suicide and voluntary active euthanasia.

Terminal sedation too is more, not less, subject to abuse in this respect since the decision to employ it can also be made either by a competent patient or the surrogate for an incompetent patient. To the extent that terminal sedation is viewed as part of standard medical practice, like forgoing life support and pain relief that may hasten death, it will be seen as an appropriate option for choice either by competent patients or by surrogates of incompetent patients, and not as requiring any special safeguards or procedures. Proponents of physician assisted suicide and voluntary active euthanasia, on the other hand, do typically insist on special safeguards and procedures to limit abuse.[13] Similarly, the Oregon statute authorizing physician assisted suicide also contains such safeguards, including restrictions that preclude some competent patients from obtaining physician assisted suicide, e.g. if they are not terminally ill.[14] This means that terminal sedation should not be preferred to physician assisted suicide and voluntary active euthanasia on the grounds that it is less subject to abuse than they are—it is not, even if it may be accepted by some who oppose physician assisted suicide and voluntary active euthanasia on this ground.

Finally, I want to note some advantages and disadvantages of terminal sedation in comparison with other forms of end of life decisions and care.[15] Terminal sedation has several advantages that bear on public policy about it. First, since the health care team is involved in carrying out the decision to employ terminal sedation, it can ensure that the decision to employ it, whether made by a competent patient or the surrogate of an incompetent patient, is fully informed and voluntary, which should include an exploration of alternative end of life care to terminal sedation, including hospice care. Second, since terminal sedation takes time to carry out it gives the health care team or surrogate and family some time to reconsider the decision after it has initially been made; once the patient has been sedated to the point of unconsciousness, he or she obviously cannot reconsider the decision unless the sedation is sufficiently lightened or removed to permit such reconsideration. Third, unlike physician assisted suicide, terminal sedation can be used when the patient is unable to act for him or herself by taking a lethal medication. Fourth, as Tännsjö argues, terminal sedation may be accepted as a compromise by patients, physicians, or policymakers who are opposed to physician assisted suicide and voluntary active euthanasia, even if the moral basis on which they do so is unsound.

Terminal sedation also has several disadvantages in comparison with other forms of end of life decisions and care. First, and as noted above, unlike physician assisted suicide and voluntary active euthanasia it can be done without the patient's consent and so in that regard is more subject to abuse. Second, it generally cannot be done at home where many terminally ill patients would prefer to spend their last days and to die. Third, there is some controversy in the anesthesia literature about whether the sedated patient might still suffer, but only be unable to report it.[16] Fourth, in a few cases terminal sedation cannot relieve all symptoms; for example, of a patient with uncontrollable bleeding from an eroding lesion or a coagulation disorder, of a patient unable to swallow from oropharyngeal cancer, or of a patient suffering refractory diarrhea from HIV-AIDS. Fifth, when patients are dying but not undergoing serious pain many physicians will find sedating them to unconsciousness medically inappropriate; if terminal sedation is considered appropriate for cases of psychological suffering as well as physical pain then it will be available to a wider class of patients. Sixth, since terminal sedation typically involves lingering in an unconscious state for days or even a couple of weeks until one dies of dehydration or malnutrition, it is incompatible with some dying patients' notions of dignity and a good death; some patients would prefer to die more rapidly and while conscious and in control once a decision for death has been made.

6. CONCLUSION

Whether terminal sedation will be the best alternative for a particular patient will depend on that patient's circumstances and values as well as on what other alternatives are medically and legally available. For the reasons that I have noted in this section, terminal sedation is not always preferable to physician assisted suicide or voluntary active euthanasia, either in individual cases or as a matter of general public policy. It should be an option available to dying patients, but so should physician assisted suicide and voluntary active euthanasia for the reasons I have also noted above. In that sense it is an alternative to physician assisted suicide and voluntary active euthanasia, as Tännsjö argues, but not always a preferable alternative, either in individual cases or as a matter of public policy. However, Tännsjö is undoubtedly correct that terminal sedation may be acceptable to many who oppose physician assisted suicide and voluntary active euthanasia, even if, as I believe, they are mistaken in their preference for terminal sedation.

Dan W. Brock
Department of Clinical Bioethics
National Institutes of Health in Washington

(The views in this chapter are the author's and do not represent any policies or positions of the US National Institute of Health, the Public Health Service, or the Department of Health and Human Services.)

NOTES

[1] Dan W. Brock, *Life and Death: Philosophical Essays in Biomedical Ethics* (Cambridge: Cambridge University Press, 1993) p. 209.
[2] Cf. Timothy E. Quill, Rebecca Dresser, and Dan W. Brock. :The Rule of Double Affect - A Critique of its Role in End of Life Decision Making,' *New England Journal of Medicine*, 337 (1997) 1768-71 and Timothy E. Quill, 'The ambiguity of clinical intentions," *New England Journal of Medicine* 329 (1993) 1039-40.
[3] Shelly Kagan, 'The Additive Fallacy," *Ethics*, 1988:5-31.
[4]. See Dan W. Brock, *Life and Death: Philosophical Essays in Biomedical Ethics* (Cambridge: Cambridge University Press, 1993).
[5] Ronald Dworkin, *Taking rights Seriously* (Cambridge MA: Harvard University Press, 1977) ch. 6.
[6] J. David Velleman, 'A right to self-termination?' *Ethics* 109 (April 1999) 606-628
[7] James F. Bresnahan, S.J. 'Observations on the rejection of physician-assisted suicide: a Roman Catholic perspective," *Christian Bioethics* 1, 3 (December 1995) 256-84.
[8] Dan W. Brock, *Life and Death: Philosophical Essays in Biomedical Ethics* (Cambridge: Cambridge University Press, 1993).
[9] Compassion in Dying v. Washington, 79 F3d 790 (9th Cir. 1996).
[10] Washington v. Glucksberg, 117 S. Ct. 2258 (1997).
[11] David Orentlicher, 'The Supreme Court and Physician-Assisted Suicide — Rejecting Assisted Suicide But Embracing Euthanasia,' *New England Journal of Medicine* 331(1991):663-67.
[12] Cf. Richard F. Ullman, Robert A. Pearlman, and K. C. Cain, 'Physicians and spouses' predictions of elderly patients' treatment preferences," *Journal of Gerontology* 43 (1988) 115-21, Tom Tomlinson, et. al., 'An empirical study of proxy consent for elderly persons," *Gerontologist* 30 (1990) 54-61, and Robert Pearlman, Richard f. Uhlmann, and Nancy Jecker, 'Spousal understanding of patient quality of life :implications for surrogate decisions," *Journal of Clinical Ethics* 3 (1992) 114-21.
[13] Charles H. Baron, Clyde Bergstresser, Dan W. Brock, et. al., "A Model State Statute to Authorize and Regulate Physician-Assisted Suicide," *Harvard Journal of Legislation*, 331(1996) 1-34.
[14] Or. Rev. Stat. 127.800(1) (Michie Supp. 1996).Ann.
[15] Timothy E. Quill, Bernard Lo, and Dan W. Brock, 'Palliative Options of Last Resort: A Comparison of Voluntarily Stopping Eating and Drinking, Terminal Sedation, Physician Assisted Suicide, and Voluntary Euthanasia' *Journal of the American Medical Association*, 278 (1997) 2099-04, reprinted as Chapter 1 of this book.
[16] J. M. Evans, "Patient's experience of awareness during general anesthesia: consciousness, awareness and pain," in M. Rosen, and J.M. Linn *General Anesthesia* (London: Butterworths, 1987) 184-192.

CHAPTER 8

LUKE GORMALLY

TERMINAL SEDATION AND THE DOCTRINE OF THE SANCTITY OF LIFE

1. INTRODUCTION

This chapter has been written in response to a request to consider whether a certain practice of terminal sedation is consistent with upholding the doctrine of the sanctity of human life; the editor believes it is consistent.

For the purposes of this symposium, terminal sedation is characterised as the medical practice of (a) sedating patients until they are comatose, and keeping them comatose until they die, in order to relieve them of the experience of conditions they find unacceptable, at the same time (b) ensuring that they are deprived of food and fluids *in order to hasten their deaths*. Terminal sedation so understood is deemed acceptable for competent patients who choose it because continued experience of conscious existence has become unacceptable and for incompetent patients about whom one reasonably believes they would have wanted it for themselves had they been in a position to make the choice.

Professor Tännsjö concludes his background paper to the symposium with the claim that "… adherents of the sanctity of life doctrine, who oppose euthanasia, … can accept a practice of terminal sedation and yet, for all that, stick to the sanctity of life doctrine and their opposition to euthanasia". How does Tännsjö get to this conclusion? The following line of thought seems to be what is offered in support:

(1) First, Tännsjö holds that "the most common argument against euthanasia has been that the practice of euthanasia must come to violate two basic principles of medical ethics: the principle of acts and omissions and the principle of double effect".

(2) However, "in most Western countries" doctors engage in intentional killing by omission and the courts accommodate the practice (witness the *Bland* case). So

Torbjörn Tännsjö (ed.), Terminal Sedation: Euthanasia in Disguise?, 81-91.

there is no continuing place for the traditional application of the doctrine of double effect in respect of responsibility for the causation of death by planned *omission*.

(3) This means that "standard thinking" about the intentional causation of death is that it is strictly prohibited only in cases in which what brings about death is a positive act aimed at causing death. 'Standard thinking' does recognise cases in which intentional causation of death by omission is wrong, but it is consequences alone which make it wrong in such cases. Since, however, the consequences of the practice of terminal sedation are acceptable, intentional killing by that element of the practice aimed at hastening death – namely, the *omission* of food and fluids – is acceptable.

I fail to see that Professor Tännsjö has anything more than the above to offer in support of his claim that acceptance of the practice of terminal sedation is compatible with upholding the doctrine of the sanctity of human life and continuing to oppose euthanasia. Manifestly his reasoning supports no such conclusion. All it serves to show is that terminal sedation is compatible with his characterisation of what he takes to be 'standard thinking' on the part of doctors and the courts. Allowing, for the sake of argument, that his characterisation does indeed identify 'standard thinking', what is one to think of the intellectual claims on us of this thinking? Not much, I suggest. Tännsjö refers to the *Bland* case as offering clear evidence of the purchase of 'standard thinking' on the courts. But what did the three Law Lords, who explicitly stated that the proposal to withhold tubefeeding from Anthony Bland was intended to hasten his death, themselves think of the distinction on which they relied between intentionally killing someone by a positive act and doing so by planned omission? It was "a distinction without a difference" according to Lord Lowry; it was a "morally and intellectually dubious distinction" according to Lord Mustill. And Lord Browne-Wilkinson confessed that "the conclusion I have reached will appear to some to be almost irrational … I find it difficult to find a moral answer".[1] What cases like *Bland* show is not the intellectual merits of Tännsjö's 'standard thinking' but rather the willingness of the judiciary to acccommodate what doctors want to do. The lesson is sociological not ethical.

To show what is compatible with the doctrine of the sanctity of human life it is necessary to begin with some exposition of the doctrine. The principle of double effect, in its application to choices which result in the causation of death,[2] cannot be understood except in the light of the doctrine of the sanctity of life. And that doctrine itself excludes any principle characterised simply as a principle of acts and omissions. Traditional moral opposition to euthanasia, grounded as it is in the doctrine of the sanctity of human life, also requires an explanation of that doctrine.

I propose, then, first to give an exposition of the doctrine of the sanctity of life (section 2), an exercise useful in itself since many critics of the doctrine fail to give a fair statement of it. For present purposes I limit myself to an explanation without attempting a full-scale defence of the doctrine. Following that explanation, it will be shown why the practice of euthanasia is incompatible with holding the doctrine of the sanctity of life (section 3). Since it will then be clear that terminal sedation, *as that term is understood in the present symposium*, is a form of euthanasia, it will be clear why terminal sedation so understood is incompatible with the doctrine of the sanctity of life (section 4). However, there is a different practice, also known as 'terminal sedation', which is compatible with the doctrine of the sanctity of life. The rationale and the limits of this practice will be explained in the final main section of the paper (section 5).

2.THE DOCTRINE OF THE SANCTITY OF HUMAN LIFE[3]

The ethical core of the doctrine of the sanctity of human life is an absolute (i.e. exceptionless) prohibition on intentionally killing another human being for reasons incompatible with justice. An absolute prohibition necessarily bears not on physical causation as such but on chosen courses of conduct,[4] i.e. courses of conduct specified by the reasons for which they were chosen. A course of conduct is indentifiable as intentional precisely by reference to the practical reasoning of the agent. Thus a course of conduct is a case of intentional killing if what results in the killing was brought about, or allowed to happen (when it might have been prevented), because an agent chose that course of conduct in order to bring about the death of another. The purpose of securing the other person's death was the *reason* for the agent's action.

The ethical core of the doctrine of the sanctity of human life is an absolute prohibition on *intentionally* killing for reasons incompatible with justice. An absolute prohition bears on what is *intentional* for two main reasons. First, because persons are most fully answerable for those courses of conduct they decide on in the light of their reasoning about their goals and the means to achieving them. At the other extreme to such fully deliberative choices are those states of affairs one brings about without intention or foresight of doing so and for which, absent culpable negligence, one is not held responsible. In between, so to speak, are states of affairs one *foresees* one will bring about but which one's reasons for acting make no part of what one seeks to achieve. (The standard example here is hastening death as a result of the use of opiates or analgesia which are solely to control symptoms. Death in this kind of case is no part of what one seeks to achieve.) Of course, if what one is seeking to achieve is a relatively unimportant good and the foreseen side-effect of one's choice involves significant harm to someone, then one may for that reason be under an obligation to refrain from that choice. But sometimes unintended harm to another is a foreseeable outcome of the pursuit of objectives one has entirely good reasons to pursue, as when a surgeon undertakes high risk surgery to save someone's life, and the surgery itself kills the person.[5]

So the absolute prohibitions of traditional morality do not concern foreseeable causations of harm. They concern intentional actions (more broadly, intentional courses of conduct) because of the fundamental importance to morality of having people never act for *reasons* which are directly contrary to the human good, and to justice in particular. The first condition of our living well and acting well towards each other is that we are clear that it is never acceptable to act for the kind of reason acting on which would amount to intentionally wronging another.

The second reason why absolute prohibitions bear on intentional actions is directly related to the first. When we act, our choices do not merely bring about states of affairs in the world; we do not just act, as the technical jargon has it, with propositional attitudes in mind. In other words, our practical reasoning is not confined to specifying *that some state of affairs should obtain*, as that my patient should be comatose and die. My practical reasoning specifies *what I am committed to doing or to being* in order to bring about the desired state of affairs; in other words, practical reasoning specifies attributional objectives: [6] for example, that I will sedate, that I will order the withholding of food and fluids, that I will overcome, say, the resistance of nursing staff to this course of conduct; and so on. My chosen commitment to these means to my ends shapes my character. In general, reasons which specify bad objectives (whether ultimate objectives, or intermediate objectives towards achieving ultimate ones), to the realisation of which I commit myself, shape a bad character, with all the implications that has for human well-being. Thus, for example, an enacted commitment to kill for a particular kind of reason contributes to shaping a dispositon to kill for that kind of reason. In contrast to intentional courses of conduct, what I do not seek to achieve, such as the side-effects of what I do, does not involve commitments which serve to shape character in this kind of way. This understanding of moral psychology (in particular the psychology of character formation) is an important part of the background to understanding the rationale of absolute prohibitions.

For traditional morality, respect for justice rests on belief in the equality in fundamental worth and dignity of every human being. Historically, belief in this fundamental dignity has not been held to be incompatible with capital punishment, i.e. the intentional killing of a criminal on the grounds that he *deserves* to die. The reason for thinking such punishment not incompatible with belief in human dignity is that the justification of capital punishment, relying as it does on notions of answerability and desert, assumes a high conception of the dignity of the criminal. Belief in the *justifiability* of capital punishment has been the main reason for interpreting the absolute prohibition of killing for reasons incompatible with justice as a prohibition on the intentional killing of the *innocent*, that is, those not guilty of certain types of grave wrong to others of a kind deserving capital punishment. (The prohibition may be more accurately and compendiously stated as being against killing otherwise than by the lawful agent of civil authority in its role of defender of the just order of life against which these kinds of wrong are committed, and as a punishment for these kinds of wrong, or as a defence against them.[7])

When we ask what *warrants* belief in the fundamental dignity of every human being, we are enquiring into those truths which underpin the ethical core of the doctrine of the sanctity of human life and form an essential part of the doctrine's substance. Historically, the underpinning has come from the Judaeo-Christian teaching that every human being is created in the image of God and is intended to find his or her fulfilment in union with God. For a variety of reasons this theological belief is no longer as influential as it used to be in the Western world. The ethically significant substance of the sanctity of life doctrine can be defended, however, by arguing that fundamental worth or dignity belongs to human beings in virtue of their *nature* as human beings rather than in virtue of the value we attach to activities in which we are able to engage because of the development of distinctive human abilities. A specific kind of dignity does indeed attach to the person who develops the intellectual and moral virtues necessary to living well as a human being. But if we recognise something distinctively valuable in the development of such abilities we should recognise that a more fundamental value belongs to the nature in virtue of which we have the capacity to develop such abilities.[8]

Those who find this thought unpersuasive, who recognise no distinctive worth and dignity attaching to human beings just because they are human, and see worth and dignity as attaching solely to the exercise of developed human abilities (in particular the ability to make choices and thereby give a particular shape to one's life), ought to recognise that their position raises a fundamental difficulty over giving a coherent account of justice. If one possesses the dignity, which entitles one to be treated in accordance with norms of justice, only if one is in a position to exercise certain developed abilities, questions arise not only about which are the relevant abilities but also about how developed they have to be. The exercise of specifying some requisite level of developed ability is inescapably arbitrary. But it is incompatible with our most fundamental intuitions about justice that *who are the subjects of justice* – that is, who are entitled to be treated justly – should be determined in an arbitrary fashion. If arbitrariness is to be avoided, then, there does seem to be no alternative to assuming entitlement to just treatment simply in virtue of our being human. Entitlement to respect for basic human rights, however, itself implies a fundamental dignity common to human beings.

It should be clear, from the above exposition of it, that the doctrine of the sanctity of human life has no place for an unqualified distinction between acts and omissions. One may aim to bring about someone's death by what one refrains from doing just as much as by what one does. Parents can deliberately kill their children by starving them to death (by a policy of failing to feed them) as well as by battering them to death. Both would be cases of intentional killing. The doctrine of the sanctity of life, prohibiting as it does the intentional killing of the innocent whether by act or omission, is simply incompatible with Professor Tännsjö's 'standard thinking'.

Euthanasia, being a species of intentional killing, can be carried out either by a positive act (such as injection of a lethal substance) or by a course of planned omissions (such as depriving a patient of food and fluids). We should turn now to a brief consideration of why euthanasia is morally impermissible.

3. THE WRONGNESS OF EUTHANASIA

The ethical core of the doctrine of the sanctity of life is the absolute prohibition on intentionally killing for reasons incompatible with justice. Concretely, this norm means that what is prohibited is the intentional killing of the innocent, namely those whose killing cannot be justified as deserved punishment.

What considerations are offered by way of justifying euthanasia? Euthanasia is the killing of a patient, usually by a doctor, in the belief that death would be a benefit (or at least no harm) to the patient because continued existence is deemed no longer worthwhile due to the patient's condition. That condition may be one of physical or mental suffering, or, in the case of some incompetent patients, it may be one of severe dementia or unconsciousness.

Some advocates of euthanasia would not wish to justify nonvoluntary euthanasia and rest their case on the claim that competent patients should have the freedom to determine the way in which their lives end. However, it should be borne in mind that it is someone else (usually a doctor) who is expected to end their lives and it is the doctor, therefore, who is answerable for the death. His justification can hardly be confined to the claim that 'X asked me to kill him'. Nor would most doctors be even tempted to think such a statement on its own could count as justifying killing. Something more has to be said about the condition of the patient, which would make plausible the judgment that the patient no longer had a worthwhile life. For a doctor could hardly begin to think himself justified in killing a patient whose life he thought worthwhile.

So even in voluntary euthanasia (and more evidently so in nonvoluntary euthanasia) the burden of justifying killing rests on the judgment that the patient no longer has a worthwhile life.

This makes clear why euthanasiast killing falls under the norm prohibiting the intentional killing of another human being for reasons incompatible with justice. Justice presupposes the recognition of a fundamental worth or dignity common to *all* human beings. A human being is a living human body and the life of that human being is the life of that body. If you assert that somebody no longer has a worthwhile life you are implying that that human being lacks worth or dignity. For to say that the ongoing life of a particular human being is not worthwhile is to deny worth to the human being, the human person, whose life it is.

Much advocacy of voluntary euthanasia serves to obscure what is required for its 'justification'; and what is required – the adverse judgment on the worthwhileness of a person's life – makes clear why it cannot be justified. Advocacy of nonvoluntary euthanasia is increasingly frank in stating that the incompetent judged eligible for euthanasia are eligible because they lack worthwhile lives.

4. TERMINAL SEDATION

The two previous sections allow me to be brief in my discussion of the practice of terminal sedation as defined for the purposes of this symposium. That practice *aims* at hastening the death of patients rendered comatose precisely by depriving them of food and fluids. And the motive for doing so is to put an end to the business of living a life "no longer worth experiencing".[9] So plainly terminal sedation is that species of intentional killing that we call euthanasia. For the reasons already explained it is radically incompatible with the moral requirements of the doctrine of the sanctity of life.

5. IS THERE A MORALLY ACCEPTABLE PRACTICE OF TERMINAL SEDATION?

The doctrine of the sanctity of life prohibits intentional killing of human beings for reasons incompatible with justice. It imposes therefore an absolute *negative* obligation, namely an obligation to *refrain* from such killing. It imposes no unqualified positive obligation to seek to prolong life in all circumstances. It is often represented as doing so, and the misrepresentation then becomes the straw man for polemics against the doctrine.

There are a number of reasons why there clearly cannot be an unqualified positive obligation to seek to prolong life:

— the first is that all of us who escape instant death will sooner or later find ourselves dying. Dying is something we need to do well but – seen from a moral or religious standpoint – can do badly. The deployment of medical technology to postpone death when a person is in the grip of the dying process can prevent that person from dying well.

— Secondly, our normal obligation to preserve our lives can sometimes be in conflict with other pressing obligations, and, if these obligations are binding, then the obligation to seek to preserve our lives will cede to them. Hence martyrs for religious truth, those who are killed in defence of their children, and officials killed in the defence of public order.

— Thirdly, no one can be obliged to take measures to preserve or prolong his life when doing so involves excessive physical, psychological or social burdens. And even if the burdens consequent on, say, medical treatment are not absolutely unbearable, if the treatment promises little in the way of increased life-span then there is hardly a warrant for considering it a matter of obligation to bear those burdens.

— Fourthly, what one's remaining life without life-prolonging medical treatment has to offer, compared with the character of one's life with treatment, may warrant the refusal of treatment. Thus, for example, an elderly woman with terminal cancer may reasonably decline chemotherapy, which offers an extra twelve months of life, because she prefers to spend the remaining six months of life, which she can otherwise expect, in the bosom of her daughter's family. No one should see anything morally problematic about such a preference when chemotherapy is likely to involve distressing physical sequelae, psychological stress and the social dislocation of regular hospitalisation.

The third and fourth points identify the reasoning behind the distinction, well-established in the sanctity of life tradition, between what are called 'ordinary and extraordinary' means of prolonging life, namely between life-prolonging treatments that are obligatory and those that are non-obligatory. The first point identifies the kind of circumstance in which, life-prolonging treatment having been excluded, there is often a need for palliative care to mitigate the symptoms of terminal illness to enable the patient to die well.

It is impossible to understand the development of the hospice movement, which, over almost half a century, has brought such dramatic improvements in palliative care, without recognising that it was inspired by a normative conception of what dying well involved. Two features in particular of this conception are relevant to our topic.

First, the movement was inspired by a firm belief in the dignity of the dying person. The varied skills available in a typical hospice are deployed to sustain in a patient a sense of his or her continuing dignity, as well as relieving physical symptoms and helping patients through psychological difficulties. Palliative care as a movement would begin to unravel if it came significantly under the influence of practitioners who thought it reasonable to hold that there were human beings who no longer had worthwhile lives. The kind of efforts which have been made in hospices on behalf of the terminally ill would come to seem redundant, since it would seem more rational to consign to oblivion and starve to death those whose lives looked as if they were no longer worthwhile. And since such a regime of terminal sedation is plainly euthanasiast, it would soon become clear that there is no good reason to take a week over killing a patient (while he or she occupies a valuable bed) when it can all be done much more expeditiously.

The second feature of the understanding of dying which has inspired the hospice movement has been a sense of the importance of securing for the patient, if at all possible, a conscious living of the experience of dying. Awareness of the proximity of death often affords to people a clear perception of the need to depart this life reconciled and at peace with God, with family and with others who have been important in their lives. And when that need is met relatives and friends can recognise the completion of a life which honours human dignity in its ending. The idea of a peaceful death consciously achieved has been part of the ideal of the hospice movement, an ideal the significance of which one can appreciate even if one attaches no sense to the specific ideal of dying at peace with God.

These two features of what I have called a normative conception of dying well explain why 'terminal sedation' has hitherto been regarded as a measure of 'last resort' in the practice of palliative care. Moreover, resort to it has been standardly and properly confined to the 'terminal phase' of dying. The 'terminal phase' is one in which "[the] patient's condition leaves no room for doubt that death is now near and is likely to occur in a matter of days".[10]

Two kinds of circumstance, in the terminal phase of dying, have been thought to justify the 'last resort' use of terminal sedation. One has been when the physical pain suffered by the patient has been extremely severe and has failed to respond to other measures to alleviate it. The other has been when the patient is suffering 'terminal anguish'. This condition is described in the following terms in a major textbook of palliative medicine:

> Terminal anguish is a tormented state of mind which relates to long-standing unresolved emotional problems and/or interpersonal conflicts, or to long-hidden unhappy memories often with guilty

> content. These problems have festered in the mind but have never been brought into the open.
>
> As long as the patient is well enough to control his/her thoughts and as long as denial can function, all appears to be well. With increasing weakness, the onset of drowsiness and inability to control thoughts, hidden matter in the unconscious is able to surface. The mental anguish manifests with restlessness, thrashing about, moaning, groaning, and even crying out. Sedation, if inadequate, only makes matters worse and nothing short of deep unconsciousness, natural or induced, provides relief. In this tormented state, when recovery is impossible and death is near, heavy sedation to give respite is merciful.
>
> The possibility of such an outcome highlights the need to make every effort to deal with psychological *skeletons in the cupboard* before the patient becomes too weak to be able to address them. A few, however, resist every attempt to share what they have been hiding.[11]

The employment of terminal sedation for terminal anguish even in the terminal phase of dying is not unproblematic. The greatest caution is surely required in deciding that a patient is beyond the reach of a word or gesture which might help to bring some peace to him. Respect for human dignity should make us reluctant to eliminate the possibility of intelligent responsiveness even when the capacity for it seems very impaired.

That said, however, the employment of terminal sedation in the terminal phase of dying, for extremely severe and intractable pain or for terminal anguish, may clearly be aimed not at the death of patients but rather at the alleviation of those conditions.

If terminal sedation is contemplated only when death is likely to occur "in a matter of days", there is no need to provide food and fluids for the dying patient. To do so would be to secure little or no benefit for the patient. Moistening of the mouth is likely to be sufficient. Conscious patients in the terminal phase of dying often decline food, while continuing to desire to be relieved of sensations of thirst.

Even if a regimen of terminal sedation in the circumstances envisaged here were slightly to hasten death (which seems unlikely), that regimen would not have been chosen precisely with a view to hastening death. If death is hastened it is an unintended side-effect of a measure adopted purely to relieve a patient of the experience of being overwhelmed by otherwise unrelievable symptoms. That regimen is therefore clearly compatible with the doctrine of the sanctity of human life.

If there are circumstances in which terminal sedation is indicated *before* death is likely to occur in a matter of days, then food and fluids should be provided. Even if the provision of food and fluids requires medical skill to establish and monitor it, it should not be categorised as medical *treatment*. It is more accurate to regard it as medical *care*. When a doctor undertakes care of a patient in circumstances in which the patient is in a position of total dependency, what is first of all owing to that patient is respect for his or her basic needs. The provision of food and fluids is an elementary expression of respect for the ongoing *life* of a human being; without them the person cannot survive. Deliberate withholding of food and fluids, when one could and should provide them, is a serious moral offence even when not intended to hasten death; though it is not as grave an offence as intentionally starving someone to death.

6. CONCLUSION

The practice of terminal sedation, as proposed for approval by Professor Tännsjö, certainly cannot be commended to palliative care physicians as consistent with respect for the sanctity of human life. It is a form of euthanasia. That does not mean that no practice of terminal sedation is possible. It should, however, be confined to exceptional circumstances in the terminal phase of dying, and innocent of *any intention* to hasten the death of patients. Outside the terminal phase of dying, if there is ever justification for it, it should be accompanied by the provision of food and fluids.[12, 13]

Luke Gormally
The Linacre Centre for Healthcare Ethics, London,
and Ave Maria School of Law, Ann Arbor, Michigan, USA

NOTES

[1] See [1993] 2 *Weekly Law Reports* at p.379 (Lord Lowry), p.399 (Lord Mustill), p.387 (Lord Browne-Wilkinson).

[2] The principle of double effect applies to many kinds of choice other than those which result in the causation of death.

[3] In the main text I shall mostly abbreviate the name of the doctrine to 'the doctrine of the sanctity of life' but it should be understood that it is always *human* life that is being referred to.

[4] I use the phrase 'chosen course(s) of conduct' to refer to deliberate omissions as well as actions.

[5] To say that there can be difficulties for an *observer* in distinguishing between what an agent intends and what he foresees does not provide a ground for calling in question the moral significance of the distinction between intention and foresight. Furthermore, any such difficulties are not such as to render the doctrine of double effect (which assumes the moral significance of the distinction between intended and foreseen consequences) inapt for employment in the legal control of medical practice. On the claim that they do render it inapt, see the response made in the Report of the House of Lords Select Committee on Medical Ethics: "Some may suggest that intention is not readily discernible. But juries are asked every day to assess intention in all sorts of cases, and could do so in respect of double effect if in a particular instance there was any reason to suspect that the doctor's primary intention was to kill the patient rather than to relieve pain and suffering. They would no doubt consider the actions of the doctor, how they compared with usual medical practice directed towards the relief of pain and distress, and all the circumstances of the case." House of Lords, Session 1993-94, *Report of the Select Committee on Medical Ethics. Volume I: Report.* London: HMSO 1994, #243 (p.50).

[6] On the importance for the understanding of morality of the distinction between propositonal and attributional objectives, see A W Müller, 'Radical Subjectivity: Morality versus Utilitarianism'. 19 (1977) *Ratio*: 115-132.

[7] It should be noted that the reason referred to here for saying capital punishment is an exception to the absolute prohibition on intentional killing for reasons incompatible with justice is intended to show only that capital punishment as a type of action is not intrinsically unjust. To take this view is compatible with objecting, for other reasons, to the use of capital punishment.

[8] For some development of this line of thought, and critique of counterpositions, see Luke Gormally (ed) *Euthanasia, Clinical Practice and the Law*. London: The Linacre Centre 1994, esp. pp.118-133.

[9] [Tännsjö, Chapter 2, this book, p. 29]

[10] R G Twycross and I Lichter, 'The terminal phase', in D Doyle, G W C Hanks and N MacDonald (eds) *Oxford Textbook of Palliative Medicine*. Oxford: Oxford University Press 1993, p.651.

[11] Twycross and Lichter, *op.cit.,* p. 659.

[12] I have discussed some of the North American advocacy of terminal sedation in Luke Gormally, 'Palliative treatment and ordinary care', in J Vial Correa and E Sgreccia (eds) *The Dignity of the Dying Person.* Vatican City: Libreria Editrice Vaticana 2000, pp.252-266.

[13] I am grateful to my colleague Dr Helen Watt and, more particularly, to Dr Mary Geach, for their comments on an earlier version of this paper.

CHAPTER 9

DANIEL CALLAHAN

TERMINAL SEDATION AND THE ARTEFACTUAL FALLACY

Why It Is a Mistake to Derive an 'Is' From an 'Ought'

1. INTRODUCTION

I want to make a two-fold argument in this chapter. First, I will try to show that an elimination of the killing/allowing-to-die distinction is a serious philosophical error. Second, I will attempt to argue that terminal sedation followed by a withdrawal of artificial nutrition and hydration is morally legitimate — but does not require the elimination of the distinction to make that case. There is another, more plausible, way to do so. Along the way I will introduce the idea of the "artefactual fallacy," deducing an "is" from an "ought."

2. KILLING AND ALLOWING TO DIE

Among most Anglo-American philosophers it has become settled dogma that there is no moral difference between killing a person directly (euthanasia) and terminating life-sustaining treatment (allowing to die)—and earlier called the difference between active and passive euthanasia. The philosopher James Rachels published an influential article in *The New England Journal of Medicine* in 1975, and it has been hard ever since to find any serious dissent, at least among philosophers (physicians seem to continue accepting the distinction).[1] The benefit of rejecting this distinction is that it more easily legitimates euthanasia and, with it, terminal sedation. If it morally acceptable, for instance, to turn off the respirator of a dying patient — the claim goes — it should be no less acceptable to kill that patient directly (and perhaps more mercifully), or to put a patient in a deep, irreversible coma — and then to

Torbjörn Tännsjö (ed.), Terminal Sedation: Euthanasia in Disguise?, 93-102.

combine that latter action with a cessation of artificial nutrition and hydration, ensuring the death of the patient.

I can most easily present my argument against eliminating the distinction by offering an interpretation of three stages of life and death, both historically and in the life of a patient.

Stage A: The healthy person and pre-modern medicine. If there is no illness or disease, the human body functions well, requiring no medical support to remain alive. It is biologically self-sufficient. When that body became sick, however, pre-modern physicians could do no more than offer some diagnostic insight or advice on healthy living, unable to intervene effectively if illness struck. If the body was to heal, it had to heal itself. While there were rules against euthanasia and assisting a patient to commit suicide, there were no rules about the use of medicine to preserve life, which was beyond its power.

Moral Rules Generated in Pre-Modern Medicine:
— diagnosis when possible
— no euthanasia or Physician-assisted suicide
— comfort to be given at all times

Stage B: The onset of a lethal disease and the provision of life-extending treatment in modern medicine. With the work of Francis Bacon and Rene Descartes in the 16th and 17th centuries, the idea of using scientific knowledge to cure illness and save life was introduced, even though it was then not actually possible to do so in an serious fashion. Medicine eventually moved on, making it possible by the mid-19th century to save and extend life, even if erratically. A new, second stage of care was introduced to the patient, accelerating by the end of that century and into the early 20th century. For the seriously ill patient hope became a realistic possibility. And when medicine historically reached that stage of its history, moral rules had to be constructed to deal with the new efficacy. I say "constructed" because neither medicine itself nor nature revealed any obvious guidance about what those rules should be. Rules were, however, devised over a period of time and are still in the process of construction as new technological possibilities and problems appear.

Moral Rules Generated:
— if life extending treatments are available for otherwise lethal diseases, they should routinely be provided to patients if not unduly burdensome: doctors "ought" to provide life-extending treatment as a general rule.
— a doctor in such circumstances who does not provide treatment that could be efficacious, or who terminates treatment already begun in a similar circumstance, is guilty of a serious moral wrong, the moral equivalent of directly killing the patient (and will be legally culpable for the omission). It was understood, however, that the patient died from the underlying illness, not the cessation of treatment, whatever the physician's culpability.

Stage C: When life-extending treatment proves futile, unduly burdensome, or is rejected by the patient. This is the most recent historical stage where standard life-extending treatment, otherwise morally required, comes to appear questionable or wrong. Problems at this stage have become exacerbated because of the power of technology to extend failing life but often without providing more than marginal medical benefit, if any, and sometimes increasing the suffering of the patient. The ordinary treatment may seem futile in further arresting the lethal disease, physically or psychologically burdensome to the patient without corresponding benefit, or simply rejected by the (competent) patient.

Moral Rules Generated:
— it is morally wrong to deny a competent patient the right to refuse treatment, even if efficacious life-extending treatment
— it is morally acceptable for treatment to be terminated on an incompetent dying patient when the treatment is futile in arresting the course of the lethal disease, or when it is physically or psychologically an undue burden
— it is morally acceptable to use means of pain relief that run the risk of killing the patient, but not when they make certain the death of the patient (a form of euthanasia)

Proposed Additional Rule
— it is morally acceptable to put a patient in a terminal, permanent coma and then to cease providing artificial nutrition and hydration without thereby incurring the charge of euthanasia.

3. THE ARTEFACTUAL FALLACY

To put that "proposed additional rule" in the context of my general argument, I want now to explicate what I will call the "artefactual fallacy," by which I mean deriving and "is" from an "ought." A failure to take account of this fallacy, I believe, lies behind the judgment that there is no moral difference between killing and allowing to die. My contention is that, when treatment is stopped, the older view is correct: the patient is actually "killed" by the underlying lethal pathology, which has been temporarily arrested, not eliminated. The lethal disease, not the cessation of treatment, is the physical cause of death (as an autopsy would show). *It was only the construction of the moral rule positing an obligation to treat that allowed us to call a violation of that rule a culpable act and to treat it "as if" it was a form of direct killing and equivalent to the [physical] "cause" of death.* When that move is made, it then becomes easy to believe that there is no moral difference between killing and allowing to die—and no causal difference as well. The artefactual fallacy has been committed, usually unknowingly.

Let me provide a background defense of those assertions. The belief that it is not possible to derive an "ought" from an "is" — called the "naturalistic fallacy" — has been long accepted in moral philosophy (though dissenters can be found).The way the world is does not tell us how we ought morally to respond to that world.

Similarly, the fact that a person is biologically dying does not tell us how we ought to medically treat that patient — which is why moral rules had to be constructed when it became possible to provide life-extending treatment.

I want to define an "artefactual fallacy" as the opposite of the naturalistic fallacy: that of deriving an "is" from an "ought." The elimination of the distinction between killing and allowing to die commits this fallacy. It forgets that the rules concerning the cessation of treatment with dying patients are socially constructed rules, that is, moral artefacts that exist because humans invented, rather then discovered, them. When, therefore, it is argued that there is no difference between killing and allowing to die, a significant fact has been forgotten: that we have decided to act "as if" allowing a patient to die is a form of physical causation equivalent to directly killing a patient. But, biologically speaking, that is not true; it is only our constructed moral rule that makes it seem true. We have, that is, come to treat the "ought" as an "is," treating the moral rule as a biological fact in the world, not a moral construct.

What is a fact is that terminally ill patients whose lives have been medically extended live longer lives than those who have not had such treatment. They thus die at a later time than would otherwise be the case. If not killed directly, they will eventually die as a result of their (temporarily arrested) lethal disease either because treatment was stopped or because even the best available treatment no longer could stay the hand of death.

If my analysis is correct, then it also becomes wrong to claim that a physician who terminates treatment is, as a common phrase has it, "hastening" the death of the patient. A physician who has saved a patient from death at time "x," medically sustaining the life of the patient but unable to eliminate the underlying lethal pathology, can not be accused of "hastening" death when, at a later elapsed time "y," he terminates life-extending treatment. The patient has already lived longer than if there had been no intervention in the first place. The overall course of dying has been slowed, and death can not meaningfully said to be hastened if treatment is terminated after a significant period of time has passed between "x" and "y."

Here is an analogy. If , as a good swimmer, I see someone drowning, the moral rule in our society is that I ought to make every possible effort to save that person. However, if I begin carrying that person into shore but eventually became unable to carry him any further, releasing him and thus allowing him to drown, I would not be accused of killing that person, even though it is the act of letting the person go that precipitates the final drowning. Nor would I be accused of "hastening" has death even if, had I tried, I might have carried him a few more meters. If, however, I deliberately let the drowning person go when I could easily have continued the rescue, then I would be held culpable for the act; and I could be treated "as if" I had killed that person. Either way, though, it is the water that biologically kills the drowning person, not my act of letting go.

Another example is relevant also. In his article James Rachels presents two cases, much cited thereafter, designed to show how the distinction between killing and allowing to die vanishes on closer examination:

"In the first [case], Smith stands to gain a large inheritance if anything should happen to his six-year-old cousin. One evening while the child is taking his bath,

Smith sneaks into the bathroom and drowns the child, then arranges things so that it will look like an accident.

In the second [case], Jones also stands to gain if anything should happen to his six-year-old cousin. Like Smith, Jones sneaks in planning to drown the child in the bath. However, just as he enters the bathroom Jones sees the child slip and hit his head, and falls face down in the water. Jones is delighted; he stands by, ready to push the child's head back under if it is necessary, but it is not necessary. With only a little thrashing about, the child drowns all by himself, 'accidentally,' and Jones watches and does nothing."

In commenting on these two cases, Rachels says that "Now Smith killed the child, whereas Jones 'merely' let the child die. This is the only difference between them. Did either man behave better, from the moral point of view? If the difference between killing and letting die were in itself a morally important matter, one should say that Jones's behavior was less reprehensible than Smith's." Rachels denies, correctly, that Jones behaved less wrongly than Smith and contends, also correctly, that Jones could not get away with the defense that, concerning the child "'I didn't kill him; I only let him die.'"

There are two problems with this argument, one bearing on the meaning of the alleged moral equivalency of the two cases, the other bearing on the larger implications of Rachels's analysis for patient care and physician responsibility. On the first issue, the reason Rachels can contend that the moral blame is equal in both cases is that we have a social rule that, if one has a moral responsibility for the welfare of another, as an adult would for a six-year-old child, then a failure to save the life of such a child from drowning would make them culpable for a failure to do so. Smith would be morally culpable for killing the child, but Jones no less so for a failure to act when he should have. But there is a difference between killing and allowing to die in these cases. In the instance of Smith directly killing the child, the child dies because of the physical action of Smith; but for that action the child would still be alive. Jones, however, is not the physical cause of death, though because of our moral rules we hold him equally responsible for the death as if he had directly killed him.

Our moral rule in effect say that, in such situations, the actual physical cause of death is irrelevant; we have declared it so by virtue of our moral rule. But it is important to remember that the physical cause of death has not been abolished; it has been overlaid by a moral rule that declares it irrelevant. If, for instance, Jones had come upon a drowning insect in the bath tub and done nothing, we would have no hesitation in declaring the water as the cause of death and excused Jones of any responsibility. There is no rule of rescue with insects; we leave the insect's fate to nature.

The attribution of moral responsibility by means of a moral rule is, then, determinative of the moral significance of the actual physical cause of death. That conclusion becomes particularly important in the medical context. The Smith-Jones cases offer only a limited insight into medical decisions: we would have no trouble declaring immoral a doctor who turned off a respirator on a competent patient who wanted to be treated and one who, finding the respirator accidentally unplugged, decided not to plug it back in. That is hardly a common medical problem, at least as rare — and simple to solve morally — as the situation of Smith and Jones.

The full moral complexity for physicians arises when they have to decide whether the general moral rule to actively treat the critically ill and dying ought, because of the patient's medical status, to be set aside in favor of treatment termination and palliative care only. The current moral rule says that may be done if there is sufficient reason to do so. The question before us is whether, if a treatment termination is judged morally acceptable and the physician so acts, it can be said that the physician has acted no differently than if he had directly killed the patient by, say, an injection. My contention is that it is in most circumstances an entirely different act. It would only be similar if, for self-interested purposes, the physician terminated otherwise efficacious treatment in order to end the life of the patient — and even in that case it would be the underlying lethal disease that would be the necessary condition for such an action to kill the patient.

The truly pervasive problem for the physician has, at it core, a great puzzle and moral dilemma. On the one hand, nature has so ordered life biologically that, sooner or later, every patient will die. Unless we think that death is a biological accident, then one can add that not only *will* every patient eventually die, but every patient must die. At best the physician (or series of physicians, as the case may be) can only forestall death, not eliminate it. We will all die at some point but neither we nor our physicians know just when that will be.

On the other hand, the medical struggle consists of giving us as much time as possible, but only as long as that can be done without an undue burden of suffering. To properly treat the patient, then, the physician must be prepared to cease medically sustaining the patient when the treatment is no longer erffective. If that is not done, the patient will eventually die anyway. But the patient, whose life has been artificially sustained, will die earlier than if that support had been continued. Yet the overriding reality is the inevitability of death. At stake is only its timing and circumstances, over which there can be some control. The moral debates have mainly turned on the issue of how much control under what circumstances to be determined by which persons.

One trend in recent decades has, seemingly, strengthened the contention that there is no moral difference between killing and allowing to die. It is that more and more lives, probably the majority, come to an end because of a decision to cease life-extending treatment. People do not, in other words, just die any more because of acts of nature but because of a decision to cease treating them, which will bring about death. It is human choice, not nature, that now ends most lives.

But that practice does not show that the killing-allowing to die distinction has been erased; the logical status of the distinction remains. But the practice might lead people to think, mistakenly, that nature no longer takes lives, only the actions of physicians. Except for the action of the physician in stopping treatment, the patient would have lived longer. Yet, in line with my earlier arguments, I would say that it is still the underlying disease that is the actual cause of death, not the physician's cessation of treatment. To change the timing, and only that, of an inevitable death from causes beyond the physician's control is not to kill the patient—even though it may appear that way to the outside observer. That observer has failed to note that we have come to think of treatment termination as the moral equivalent of killing, as if the physician was directly ending the patient's life. But it is only our socially constructed moral rule that treatment may, under certain circumstances, be terminated if various moral conditions have been met. The observer, in short, has committed the artefactual fallacy.

The artefactual fallacy then (a) treats our constructed moral rules as if they are biological facts, and (b) then equates them with direct killing, and (c) then holds that, morally speaking, there is no difference between killing and allowing to die. An "is" has thereby been derived from an "ought."

4. TERMINAL SEDATION

What has all that to do with terminal sedation? Two things. The first is that there is now a debate about what rules ought to be devised for the employment of terminal sedation. Is it to be morally acceptable to sedate someone into a permanent coma? And, if that is done, is it further acceptable to cease all artificial nutrition and hydration, thereby ensuring the death of the patient more quickly ("hastening" it) than would otherwise be the case? There appears to be much more debate in response to the second question; terminal sedation now seems generally accepted.

The sticking point is that the cessation of artificial nutrition and hydration is that it looks like a case of euthanasia, that is, directly intending the death of the patient and taking steps certain to bring about that death. If, however, one believe there is no difference between killing and allowing to die, then no real moral dilemma exists: it is as acceptable to stop artificial nutrition and hydration, killing the patient, as to stop any other medical intervention. For those who hold that position, the cessation of artificial nutrition and hydration directly kills the patient, but that is of no moral importance, differing not at all from the cessation of any other interventionist treatment.

I want to approach the problem in a different way, not resting the outcome on a supposed identity between killing and allowing to die—not, that is, committing the artefactual fallacy. It is possible to make a case that it is morally acceptable to put someone in a permanent coma and then to terminate artificial nutrition and hydration without committing that fallacy.

I begin with a proposed new rule pertinent to the use of technology to sustain and extend a life burdened by a lethal disease:

— There is no moral obligation to use a life-extending technology that would never have been invented solely to sustain the life of a dying patient, much less a patient in a terminal coma. In medicine, many technologies are invented for one purpose but eventually come to be used for other and additional purposes. Artificial hydration and nutrition were originally developed for a specific short-term purpose, that of helping a patient to recover from surgery, supplying artificially for a brief period the food and water that could not be given by mouth. By the 1970s, however, improved tubing and other methods of artificial nutrition and hydration made it possible to sustain patients indefinitely on artificial nutrition and hydration. But the aim of the improvements was still to help a patient weather a critical illness for a short time—not to find a way to keep an irreversibly comatose alive indefinitely or to permanently sustain patients put into a coma to relieve their pain. The fact that this can be done with such patients does not entail that there is an obligation to do so. I define technological captivity as the belief that, if a technological is available, it ought to be used. That seems to me wrong, and as a broad proposition, most people would probably reject it. The fact that artificial nutrition and hydration is technically available ought not to require its use on terminal patients, especially when it was not designed for that purpose. It is hard to imagine that anyone would have tried to invent technologically sophisticated artificial nutrition and hydration technology exclusively to prolong the lives of those in irreversible comas, and especially if the patient was so treated to relieve pain and suffering.

In any event, there has for some time been a debate about whether the withdrawal of artificial nutrition and hydration from a terminal patient should be considered direct killing, euthanasia. Such a withdrawal will end the life of a patient and it is definitively known that it will do so. Yet there are two reasons for not classifying it as euthanasia. The first reason is that any rule requiring artificial nutrition and hydration is a constructed rule, one historically devised to deal with patients in a temporary, not permanent, coma. There is no known medical or health benefit or moral reason to extend that rule to those who are terminally ill and in a permanent coma. There is no benefit in such an extension, and a moral rule requiring it would be a mistake. The ultimate technological captivity is to classify any failure to use technology to sustain life as equivalent to directly killing a patient. But the human body eventually dies, and the only serious moral question is whether the patient's condition and prognosis warrants an effort to give the patient some additional time, but always a finite amount of time.

The second reason is that an inability to take water or nutrition by mouth, whether for physical or psychological reasons, is a classic sequelae of the dying process. It is in that sense "natural," that is, a biologically predictable part of dying. To allow that process to go forward, when it is understood that the patient is dying, is an instance of "allowing to die," not of direct killing. It is not the termination of artificial nutrition and hydration that kills the patient, but the underlying lethal condition. And it was that condition which morally permitted putting the patient into a permanent coma. Except for the constructed (and sometimes controverted) rule about an obligation to provide artificial nutrition and hydration even to the permanently comatose, there would be no moral rule to provide it. And there should be no such rule. To call the cessation of artificial nutrition and hydration euthanasia would be to commit the artefactual fallacy.

In sum, it is not necessary to reject the distinction between killing and allowing to die in order to morally justify the cessation of artificial nutrition and hydration for a patient who has been terminally sedated for a good reason (to relieve suffering). At the same time, it would be wrong to deliberately kill such a patient by, say, a lethal injection; in that case the injection would be the physical cause of death, not the underlying pathology. If a termination of artificial nutrition and hydration will lead in a short time to the death of a comatose patient, and with no accompanying pain or suffering, there would be no need for, or justification for, a faster death by euthanasia.

Daniel Callahan
The Hastings Center
New York

NOTES

[1] James Rachels, "Active and Passive Euthanasia," *The New England Journal of Medicine* 292:2 (January 9, 1975), pp. 78-80; for an interesting dissent (which seems to have made little difference with most philosophers), see Tom L. Beauchamp, "A Reply to Rachels on Active and Passive Euthanasia," in Tom L. Beauchamp and Seymour Perlin, eds., *Ethical Issues in Death and Dying* (Englewood Cliffs, N.J.: Prentice-Hall, 1978), pp. 246-258.

CHAPTER 10

JOHANNES J. M. VAN DELDEN

TERMINAL SEDATION: DIFFERENT PRACTICES, DIFFERENT EVALUATIONS

1. INTRODUCTION

One of the classical themes of medical ethics are end-of-life decisions. The introduction, some time during the nineteenth century, of techniques that gave physicians the possibility to intervene in the way that patients were dying, lead to the question under what conditions the use of these techniques would be morally acceptable. The relatively long history of the subject might give rise to the thought that by now everything will have been said about it. This, however, is not true. Again and again new developments in the practice of medicine give rise to new discussions about the ethical aspects of end-of-life decisions by physicians. One of the issues that are heavily debated at this moment, at least in the Netherlands, is terminal sedation.

One of the characteristics of the debate about terminal sedation undoubtedly is that it is a very confused one: people disagree about the meaning of the term, the appropriateness of it and, of course, about the conditions under which it would be morally justified. As a matter of fact these discussions are deeply connected: often a discussion about terms is a discussion about norms. To call the sedation of a patient 'deep sedation' rather than 'terminal sedation' has implications for the way we evaluate these actions. In this chapter I will try to clarify the discussion about terminal sedation by means of a medicine based ethics (Van Delden 2003). I will draw on the data from the empirical research of end-of-life decisions in the Netherlands and on my own experience as a nursing home physician to see what morally relevant differences exist in that part of medical practice which is referred to as terminal sedation. In doing so, I will try to answer the question in the title of this volume: is terminal sedation a form of euthanasia in disguise?

Torbjörn Tännsjö (ed.), Terminal Sedation: Euthanasia in Disguise?, 103-113.

2. TERMINAL SEDATION IN THE NETHERLANDS

By now three nation-wide empirical studies of end-of-life decisions have been performed in the Netherlands (Van der Maas 1991, Van der Maas 1996, Onwuteaka 2003). The issue of terminal sedation was not explicitly addressed in the first two studies, but it was in the most recent one. That study concerned the year 2001 and was published in 2003. It consisted, as did the previous studies, of different sub-studies. One was an interview study with physicians using a semi-structured questionnaire and a second one a survey using a written questionnaire which was sent to the attending physicians who had filled in the death certificates of deceased patients. In this study the issue of terminal sedation was addressed both during interviews and in the death certificate study. In 2001 the respondents of the interviews were 410 physicians (general physicians, medical specialists and nursing home physicians); the death certificate study concerned 5617 deceased patients.

In both sub-studies the subject was introduced by asking whether the physician had kept his patient continuously sedated till death occurred by means of medications such as barbiturates or benzodiazepines. If the physician answered affirmatively, he was asked whether in that case food and fluids were withheld or not. For the sake of consistency with the other chapters in this book I will use the term 'terminal sedation' only for those cases in which also the second question was answered affirmatively, that is only if patients were kept in a coma continuously and foods and fluids were withheld or withdrawn.

The results of the 2001 study concerning terminal sedation have so far only been published in Dutch (Van der Wal, 2003). I have taken my description of the situation in the Netherlands from this report.

In the interviews it appeared that in 10% of all deaths (the total number of deaths in 2001 being approximately 140.000) terminal sedation had been applied. The death certificate study yielded a much lower incidence: 3,9%, which is surprising because for all other end-of-life decisions the two sub-studies yielded very similar results. The difference may be explained by the fact that, due to a routing problem in the death certificate questionnaire, the question about terminal sedation was only asked when respondents had made another end-of-life decision as well. Thus, the answers to the question about terminal sedation is missing in a number of cases. Although the incidence mentioned here for the death certificate study is corrected for the missing data, one cannot rule out that it is an underestimation of real practice. Whatever the explanation of the difference in incidence may be, one conclusion can be drawn: terminal sedation occurs quite often in medical practice, at least in the Netherlands. This holds true for general practitioners, medical specialists and nursing home physicians. A considerable number of all these groups of physicians had applied terminal sedation at least once in 2000 or 2001. The exact percentages were: 33% for general practitioners, 44% for medical specialists and 65% for nursing home physicians. We can therefore conclude that, at least in the Netherlands, the practice of terminal sedation does not confine itself to hospitals.

The study report also describes the characteristics of the patients who were terminally sedated. Of those, 50% were female, 50% male. 23% were younger than

65 years, 43% were between 65 and 79 years of age and 34% were over 80 years. Cancer was the most prominent diagnosis (66%) among patients, with cardiovascular disease in second place (17%). Most patients were close to death, even without terminal sedation. The physicians interviewed were asked to give an estimation of the time the patient would still have lived if the decision to apply terminal sedation had not been taken. According to the physicians the estimated amount of life-shortening was less than 24 hours in 42%, 1-7 days in 32%, 1-4 weeks in 21% and more than one month in 5%.

The interviewed physicians were also asked to describe the decision-making process. It appeared that the decision was discussed with the patient in 31% of cases and with the patient's family in the other 69%. The decisions to sedate deeply and to withhold fluids and food were taken at the same time in 85% of cases. Asked for their reasons for terminal sedation physicians answered that it was pain in 54% of cases, uneasiness in 43% and dyspnoea in 30%. Here, more than one answer could be given.

From a moral point of view it might be interesting to know what the physician intended to do when he took the decision. In the interview the physician could indicate whether he had acted 'taking into account the probability or certainty that the end of life was hastened', or 'in part with the purpose of hastening the end of life' or 'with the explicit purpose of hastening the end of life'. The first phrase was coined to capture a situation which could be justified by invoking the principle of double effect. In 56% of cases the physicians indicated that they had had no intention to hasten the end of the patient's life. In 28% it had been partly their intention to shorten life, and in 16% this had been their explicit intention.

It is also interesting to see what drugs were used. This may provide a validity check: did the respondent describe the same thing as the interviewer had in mind? The interviews showed that benzodiazepines (mostly midazolam) were used in 31% of the cases, and morphine in 33%. The combination of these two types of drugs was used in another 33% of the cases.

3. INTERPRETATION OF THE DATA

As was mentioned above the third nation-wide study was the first study to report on terminal sedation as a separate end-of-life decision in the Netherlands. It should be noted that this term ('end-of-life decision') is used as an umbrella-concept for a diversity of decisions ranging from non-treatment decisions, alleviation of pain and symptoms to active ending of life. Thus, although euthanasia is considered to be one type of end-of-life decisions, not all end-of-life decisions are forms of euthanasia. Moreover, it should be pointed out that in the Netherlands euthanasia implies active ending of life on the explicit request of the patient. What does the description of the cases of terminal sedation mean? Are these new cases to be added to the total incidence of all end-of-life decisions or not? The authors of the 2001 report do not think so, and I agree.

Here the death certificate study is helpful, since in this study a case is described both by a categorisation in one of the types of end-of-life decisions and by the question about terminal sedation. By studying the relation between these two descriptions the authors could break down the total incidence of 3,9% for terminal sedation. When asked to describe the most important end-of-life decision in these cases, the respondents indicated alleviation of pain and symptoms in 1,9%, non-treatment decisions in 1,5% and active and intentionally ending of life in 0,6%. The latter category consisted of 0,5% on the request of the patient and 0,1% without any explicit request. We can therefore conclude that the incidence of terminal sedation does not need to be added to the total incidence of end-of-life decisions (in 2001: 43,8% of all deaths).

These data also shed light on a separate issue namely that of the reporting procedure for euthanasia. One of the disappointing results of the 2001 study was the finding that the percentage of reported euthanasia cases had risen only from 41% in 1995 to 54% in 2001. In spite of a new law and a reporting procedure involving a committee of which a medical doctor is a member (Van Delden 2004), still almost half of all cases appear not to be reported in the appropriate way. The data given above about terminal sedation suggest at least one explanation for this apparent failure to comply with the rules.

Imagine a case in which a patient asks to be put to sleep in order to relieve his suffering. Let us assume that the physician decides to apply terminal sedation. Not only is that what he does, it is also how he describes his act to himself. With 'describes' I mean that that is how the physician labels what he does himself. He will then think that the rules for euthanasia and for the reporting of such cases are not applicable, because he believes that he has decided to apply terminal sedation on the request of the patient, not euthanasia. Therefore, he will surely not report the case to the euthanasia assessment committees. In the study, however, the researchers did not ask for the labels used by the physician, but studied separate elements of the decision: what did the doctor do, was it on the request of the patient, what was the physician's intention? If a case like the one I just sketched would be studied by means of the questionnaire it might well be that the physician would indicate that his decision was made with the explicit intention to hasten the end of life. Since also drugs were administered, the investigators would surely categorise this case as euthanasia. This, however, would lead to a situation in which the case would be counted in the denominator, but not in the numerator. The consequence is that the proportion of reported euthanasia cases drops.

The next question of course is, who was right from an ethical point of view? The physician who thought he was not actively ending life or the researchers who thought he was? This brings us to the central question of this volume: is terminal sedation euthanasia in disguise?

4. MORAL EVALUATION OF TERMINAL SEDATION

Let me start with what will turn out to be the final sentence of this chapter: "The answer to the question whether terminal sedation is euthanasia in disguise must be:

that depends!" I think that both those who claim that all cases of terminal sedation should be regarded as euthanasia and those who claim that none of them should be, are wrong. All of them overlook morally relevant differences, I believe. Now let me elaborate on these morally relevant differences. I will do so by describing situations that I have encountered both in medical practice and in my work as a member of one of the five assessment committees for euthanasia.

Suppose Ms A is a patient who is dying. She has a life-expectancy of a few days only and has lost consciousness. Unfortunately, she suffers from a terminal delirium which causes uneasiness. The situation is discussed with the three daughters of Ms A. They agree with the treating physician that their mother is suffering. Ms A. is therefore treated with a sedative and with an anti-psychotic drug. This treatment restores a peaceful dying; Ms A dies a few days later. No one ever *decided* to withhold fluids and food, because it did not even occur to anyone to start doing that. Why would anyone want to disturb these final days with needles and tubes? Yet I submit that this case would come under the description of terminal sedation, used in this volume: the patient is sedated until death follows and did not receive food or fluids.

Here is a second case: Mr B. He has a much longer life expectancy (say months) than did Ms A. He suffers from cancer and suffers deeply. Opioids have not been able to relieve his suffering (pain and restlessness) and only resulted in him getting acquainted with all the side-effects of these drugs. His physician has consulted the palliative care team of the local academic hospital (Francke 2003) which sees no alternative but to try sedation. The physician discusses this option with the patient, who agrees to it. They both feel that if sedation is only acquired at a level at which Mr B looses consciousness, artificial foods and fluids should not be administered. They feel that this would only prolong suffering, without any prospect of amelioration.

Ms C is the third patient. She has cancer and does not want to live through all the degrading (her words!) stages of that disease, so she wants euthanasia. She has only one problem with it: everything goes so quickly and therefore seems so unnatural. She would prefer a continuous drip which would cause her death over a couple of days. She suggests to her physician that terminal sedation might just do that.

Before we turn to the moral evaluation I should say that I think that a considerable proportion of the cases of terminal sedation described in the Dutch 2001 study are similar to the case of Ms A. Most of these patients were close to death, even without terminal sedation. The estimated amount of life-shortening was less than one week in 74% of the cases. This in part also explains why the decision was discussed with the family rather than with the patient in 69% of the cases: the patient had lost consciousness at the time of decision-making.

I think these different cases provide a first view of some morally relevant differences. In the case of Ms A no shortening of life resulted from applying terminal sedation, since one can assume that it takes about one week to die from dehydration. Moreover, the physician provided proportional medical care: Ms A was not comfortable in her final days. If a medical treatment is started for a good reason and was performed in the professionally right way (*lege artis*), then the patient dies of natural causes, *id est* his underlying disease. In this case terminal sedation surely did not amount to euthanasia. This is not altered by the fact that food and fluids were withheld. Professors Quill, Lo and Brock in their famous article (Capt. 1 of this volume) claim that it is "implausible to claim that death is unintended when a patient [...] is sedated [...] and fluids and nutrition are withheld, making death certain" (Quill 1997), but they must have overlooked cases like Ms A. In her case the withholding of fluids did not make death certain. It was the disease that did that. To conclude that withholding fluids also in these cases shows the real intention of the physician (supposedly to kill the patient) is certainly implausible.

The case of Mr. B is different. His life is surely shortened by withholding fluids and food from him, since he still had a life-expectancy of months. On the other hand terminal sedation in this case was a last resort for his extreme suffering. Therefore one could still say that terminal sedation was proportional medical care. If this is correct, then the same rule is applicable as in the case of Ms A. One could still hold that the cause of the death of Mr B is his underlying disease.

But am I not missing the point? Should I not have stressed the fact that the physicians both in the case of Ms A and in the one of Mr B would have indicated that they only applied terminal sedation 'taking into account the probability or certainty that the end of life was hastened'? Thus, is it not the doctrine of double effect that saves both cases, rather than the fact that terminal sedation in both cases can be described as proportional medical care? Perhaps I missed the fact that proportionality in itself is one of the conditions for applying the doctrine of double effect in the first place (Kamm 1991)!

Obviously, rather than missing the point, I explicitly left it out. I did so because I think that the principle of proportionality can be applied independently of the doctrine of double effect. Moreover, I think the intention of the physician is rather irrelevant in these cases. First of all, I am not sure both physicians would indeed indicate that they only took life-shortening into account. At the end of the life of a patient a situation may arise in which the patient is clearly not comfortable. Then the physician may hope that his patient dies a peaceful death and at the same time may hope that he will die soon. Terminal sedation might serve both these goals. Therefore, the physicians in cases A and B might also have answered the question of their intention with 'partly' or 'explicitly with the intention to hasten the end of life'. It is even not certain that different answers would reflect real differences. Intentions are to a large extent reconstructions of what one felt at the time of decision-making and they are hard to verify (Quill 1993).

My point, however, is that I do not think that this matters at all. If what the physician did was the right palliative measure and if it was therefore perfectly understandable that he hoped for the end to come, why hold it against him that he acted accordingly? We may assume that he did not intend to hasten the end of life, but even if we were wrong, we would evaluate his behaviour in the same way because his actions fulfilled the criteria of proportionality and subsidiarity: the bad consequence of his act was the lesser evil which could not have been avoided by acting in a different way.

To clarify my point let us suppose that there was another case just like Ms A. Let us call her Ms A'. Suppose that the physician in the case of Ms A had no intention to hasten the end of life, while the physician in the case of Ms A' did. All the rest is the same: the same life-expectancy, medical situation etc. Should we then say that the physician of Ms A only provided good palliative care, while the one of Ms A' committed active killing? I think it would be unreasonable to treat these physicians differently on the grounds of what they had in mind only, if what they did was exactly the same, with the same result. It would even be more unreasonable to do so since in neither of these cases the end of life was hastened by sedating the patient. But even if this causal connection is present, as in the case of Mr B I still do not think what happened could be described as euthanasia (or killing) as long as the case fulfils the conditions of proportionality and subsidiarity.

The next point of evaluation should be the difference in the medical situation between Ms A and Mr B. I think terminal sedation in the case of Mr B is a last resort. The case of Ms A., however, will be easily recognised as relatively normal by physicians who often treat terminal patients. We should then ask what the moral relevance of this difference is. I think there is none. In both cases the physician provided normal medical care. I use the phrase 'normal medical care' for treatments that are acknowledged by the medical community as part of the medical professional standard. The fact that some of these treatments will be used only very infrequently does not conflict with the fact that these treatments fall within the range of normal medical care.

There is at least one other issue that is different between the case of Ms A and the one of Mr. B. Ms. A was unconscious at the time of deciding and therefore the decision was discussed with her family who acted as her representative.[1] The decision to apply terminal sedation in the case of Mr. B., however, was discussed with himself. How much weight does this carry in a moral evaluation?

Obviously, it is preferable to start terminal sedation only after having obtained informed consent by the patient. In that way one not only pays respect to the patient's autonomy but discussing the situation with the patient also enables the physician to be better informed about the amount of suffering and thus about the necessity of using terminal sedation. This, however, does not mean that only competent patients could be terminally sedated. To claim that would be paradoxical, at least from a clinical point of view. This is so, because terminal sedation will be medically indicated especially in situations where the condition of the patient if not precludes competency then at least threatens it profoundly. People in severe pain or distress want only one thing: to get rid of the pain and the distress. To speak of obtaining informed consent in such a situation seems somewhat grotesque.

But what about incompetent patients? Should we never use terminal sedation in those cases? I submit that this would be unacceptable, because it would mean that incompetent patients would be left to suffer needlessly. To use terminal sedation in case of an incompetent patient is morally defensible if four conditions are fulfilled. It has to meet the criteria of proportionality and of subsidiarity, and the representative of the patient has to agree with it. In addition to this the patient must not have indicated before (e.g. by means of an advance directive) that he would never want to be terminally sedated. In my view, the moral acceptability of terminally sedating an incompetent patient rests not on the fact that the patient would surely have wanted this to be done as the editor of this volume seems to imply in his introduction. Rather, it is the fact that this is clearly in his best interests. This, of course, implies a form of paternalism, which is justified, I submit, by the fact that major benefit is provided in a situation where refraining from terminal sedation would certainly not result in a situation in which the patient would be able to use his right to self-determination. Moreover, I think predictive consent theories as the one professor Tännsjö seems to defend, run the risk of justifying more paternalism than is desirable. We often accept risks which rational agents would have to reject. I agree with Beauchamp and Childress who say that 'autonomy-based justifications should be kept at arm's length from paternalism' (Beauchamp 1994:281).

I have been silent about Ms. C so far. Suppose her physician complied with her request. Would he then be doing something different from granting a request for euthanasia? I think not. In that case there would not be a relation between the 'treatment' chosen and the medical condition of Ms C. Of course I am not claiming that the condition of Ms C is without any medical problem, my only claim is that sedating her is not the medical answer to her problem. That is why I put quotation marks around the word 'treatment' because I do not think the act actually deserves that label. Terminal sedation in this case was used as a technique to (intentionally) cause death. Here the physician from a medical point of view could have acted differently, but did not. There may have been good reasons for him to do so, but these reasons need to be evaluated. In the Netherlands we have assessment committees to do that. The case of Ms C should have been noted as a case of euthanasia.

To call this case a case of euthanasia of course does not complete the moral evaluation of the case. In the Netherlands a case of euthanasia is only accepted when the physician meets the criteria for due care. These criteria are as follows:

The doctor must

— be satisfied that the patient's request is voluntary and well considered;

— be satisfied that the patient's suffering is unbearable and that there is no prospect of improvement;

— have informed the patient of his or her situation and further prognosis;

— have come to the conclusion, together with the patient, that there is no other reasonable alternative;

— have consulted at least one other independent physician, who must have seen the patient and stated in writing that the attending physician has fulfilled the criteria listed in the previous four points;

— have exercised due medical care and attention in terminating the patient's life.

In other countries (with the exception of Belgium) such criteria do not exist. If in these countries the practice of euthanasia is forbidden, then that rule should, for reasons of consistency, also forbid terminal sedation in cases like that of Ms C.

5. CONCLUSION

In this volume professor Tännsjö takes the position that terminal sedation might serve as a compromise in the euthanasia discussion. Adherents of euthanasia should, according to him, accept terminal sedation because it makes death easier for those patients who want to die. And adherents of the sanctity of life doctrine should accept it because terminal sedation makes it possible to hold on to this doctrine, while at the same time being sensitive to the suffering of patients. I am afraid I have to disagree. Terminal sedation under certain conditions, especially when this treatment is the medical answer to a medical problem, has nothing to do with euthanasia and will therefore fall outside the range of things that the proponents of euthanasia are pleading for. They are not campaigning for adequate medical care (although they will certainly welcome it) but for the right of the patient to ask for death.

Under different conditions, especially when terminal sedation is only a technique to bring about death on request, it is morally equivalent to euthanasia, not just a in-between compromise position. Obviously, this will be unacceptable to adherents of the doctrine of the sanctity of life. Therefore, the answer to the question whether terminal sedation is euthanasia in disguise must be: that depends!

NOTES

[1] In the Netherlands the children of an incompetent patient have the legal authority to act as representatives.

REFERENCES

Beauchamp TL, Childress J. Principles of Biomedical ethics. 4th edition. New York: Oxford University press, 1994.

Delden JJM van. Medicine based ethics. Utrecht: Utrecht university, 2003.

Delden JJM van, Visser JJF, Borst-Eilers E. Thirty years of experience with euthanasia in the Netherlands: focussing on the patient as a person. In: Quill T, Battin M (ed). *to appear in 2004.*

Francke AL. Palliative care for terminally ill patients in the Netherlands. The Hague: Ministry of Health, 2003.

Kamm F. The doctrine of double effect. J Med Philos. 1991; 16: 571-585.

Maas PJ van der, Delden JJM van, Pijnenborg L, Looman CWN. Euthanasia and other medical decisions concerning the end of life. *Lancet* 1991; 338: 669-74.

Maas PJ van der, Wal G van der, Haverkate I, et al. Euthanasia, physician assisted suicide and other medical practices involving the end of life in the Netherlands, 1990-1995. N Engl J Med 1996; 335: 1699-705.

Onwuteaka-Philipsen BD, Heide A van der, Koper D et al. Euthanasia and other end-of-life decisions in the Netherlands in 1990, 1995, and 2001. *Lancet* 2003; 362: 395-99

Quill TE. The ambiguity of clinical intentions. *N Engl J Med* 1993; 329: 1039-40.

Quill TE, Lo B, Brock DW. Palliative Options of last resort.. *JAMA* 287;1997:2099-2104; here reprinted as Chapter 1 of this volume.

Wal G van der, Heide A van der, Onwuteaka BD, Maas PJ van der. Medische besluitvorming aan het einde van het leven. Utrecht: de Tijdstroom, 2003.

CHAPTER 11

TORBJÖRN TÄNNSJÖ

THE SANCTITY OF LIFE AND THE ACTIVE/PASSIVE DISTINCTION

A Final Reflection

1. INTRODUCTION

It has turned out in this anthology to be a controversial question how the Sanctity-of-Life Doctrine is best interpreted with respect to the active/passive distinction. Of course, there are two aspects to this question, one descriptive and one methodological. The former deals with how the adherents of the doctrine are to be understood, when they defend it. Do they rely on the active/passive distinction or not? The second has to do with normative plausibility. Will the Sanctity-of-Life Doctrine be more or less plausible, if it is taken to rely on the active/passive distinction? It is to the latter question I turn here.

Some have believed that the distinction as such makes no sense. If they are right, then the adherents of the Sanctity-of-Life Doctrine had better not rely on it. However, the distinction does make sense. This is the claim I made in Chapter 2 of this book. It has been questioned by both Dan Brock and Helga Kuhse. Let me therefore, in these concluding reflections, try to repeat and strengthen my argument to the effect that the distinction does make sense, and to the effect that, if it, together with the doctrine of double effect, is incorporated in the Sanctity-of-Life Doctrine, then the doctrine will turn out as more plausible than it has sometimes been considered to be.

Torbjörn Tännsjö (ed.), Terminal Sedation: Euthanasia in Disguise?, 115-125.

2. ACTIVE/PASSIVE – A RELATIONAL NOTION

The reason that many have believed that the active/passive distinction makes no sense is the correct observation that all concrete actions are active under some description of them. This means, as far as I can see, that we must give up the distinction between acts and omissions. However, this does not mean that we should jettison the active/passive distinction. It only means that we should conceive of this distinction as relative. At least some *kinds* of actions allow that we sort instances of them into the active or passive category, relative to the kind in question. And killing, or hastening death, is indeed an example of this. There are clear cut cases of active killing, and there are clear cut cases of passive killing (of allowing nature to take its course).

No *criterion* can perhaps be formulated here, I have noted, but no criterion is really needed, I insist in saying. Our linguistic intuitions are clear enough. In many particular cases we can say of an act of killing, not if it is 'active' or 'passive' in any absolute sense, but *qua* 'killing' we can say whether it is active or passive, and we can even state our reasons for this assessment (although these reasons cannot always take a general form).

Note the similarity here with a position known as 'particularism' in ethics. We may find particularism wanting for some reason or other (I have argued against particularism[1]) but there is no denying that it is a consistent position. According to particularism actions are right or wrong, and we are often capable of saying of a particular action if it is right or wrong, but there is no general criterion available, with which we can make the assessment in question. Now, if particularism is a viable position in ethics, it should not come as a shock that it may also have some application in linguistics.

Which are our linguistic intuitions with respect to active and passive killing, then? Not to feed a patient, who, as a consequence, starves to death, is to kill passively, while injecting an opioid, which kills the patient, is to kill actively. Of course, there may be cases where our linguistic intuitions fail (some of) us. Here is one, given by Dan Brock and referred to by him in Chapter 8 of this book:

> Consider the case of a patient terminally ill with ALS disease. She is completely respirator dependent with no hope of ever being weaned. She is unquestionably competent but finds her condition intolerable and persistently requests to be removed from the respirator and allowed to die. Most people and physicians would agree that the patient's physician should respect the patient's wishes and remove her from the respirator, though this will certainly cause the patient's death. The common understanding is that the physician thereby allows the patient to die. But is that correct?
>
> Suppose the patient has a greedy and hostile son who mistakenly believes that his mother will never decide to stop her life-sustaining treatment ... Afraid that his inheritance will be dissipated by a long

> and expensive hospitalization, he enters his mother's room while she is sedated, extubates her, and she dies. Shortly thereafter the medical staff discovers what he has done and confronts the son. He replies, 'I didn't kill her, I merely allowed her to die. It was her ALS disease that caused her death.' I think this would rightly be dismissed as transparent sophistry — the son went into his mother's room and deliberately killed her. But, of course, the son performed just the same physical actions, did just the same thing, that the physician would have done. If that is so, then doesn't the physician also kill the patient when he extubates her?[2]

What are we to say about these two cases? It seems clear, in the first case, that this is a case of passive killing. Furthermore, it seems to be a case of legitimate passive killing (of allowing nature to take its course). What about the latter case? My intuition is that this too is a case of passive killing, but certainly a case of illegitimate passive killing. From a moral point of view, it may very well be as bad as active killing, and yet it is passive. This does not mean, however, that the distinction is not real (we have learnt that from Daniel Callahan's contribution to this book).

However, some may have a different intuition. They may feel that, in the latter case, we are confronted with active killing. If they do, I think this may have something to do with the fact that the son, a third party, interferes with what the doctor does. When the doctor disconnects the patient from the ventilator, having first put the patient there, we may see this as an act of capitulation with respect to a natural course of events. Moreover, the patient would not have been on the ventilator in the first place, if it had not been for the intervention by the doctor. However, when a third party (the son) intrudes and stops the doctor from saving the life of the patient, some may conceive of this as active killing. My intuition is still that we are here dealing with a case of passive killing.

My intuition is no different when it comes to Helga Kuhse's terrible example:

> Imagine that Frieda has been given a kitten for Christmas. She puts it in box and deliberately leaves it there, without food and water, until it dies. I think it would be a mistake to say that Frieda has merely allowed the kitten to die. She has killed the kitten: by putting the kitten in the box and leaving it there until it is dead, she has not merely allowed nature to take its course, but has *initiated* a course of events that will lead to death.[3]

Here it seems to me that Frieda actively, and certainly in a cruel and immoral way, by putting the kitten in the box, *robs the kitten of its liberty*; equally, cruelly and immorally, by leaving the kitten in the box, she allows the kitten to starve to death, i.e., she passively *kills* the kitten (she *allows the kitten to die*). At least this seems fair to say if, had she provided the kitten with food and water, it would have stayed alive. It would have been better, I hasten to add, if, painlessly and actively, she had killed the kitten. Here the adherents of the Sanctity-of-Life Doctrine may well concur, since this doctrine doesn't apply to felines — it applies only to (innocent) human beings.

Obviously, Kuhse does not share my intuition. Some may agree with Kuhse, others with me. I admit that the case is not quite easy to solve. So perhaps we have here come across a case that is genuinely not (inter-subjectively) decidable.

Be that as it may, there is no need to give up the distinction only because there may exist some vague cases, cases where we are hard put to say whether they constitute active or passive killing. The possible existence of such cases means only that, to the extent that the active/passive distinction plays a crucial moral role, we will end up with cases that are, even from a moral point of view, somewhat indeterminate. Who said that all moral cases must have a determinate solution?

3. THE MORAL IMPORTANCE OF THE DISTINCTION

Now, is the distinction between active and passive killing of moral importance? I have noted that, in medical practice in modern Western countries, the distinction between actively and passively killing as such is of no direct importance. For in most Western countries, even active killing of severely ill patients is legally tolerated. I think here of cases where patients are given sedative medication or opioids in a manner that hastens death. *This* is clearly a case of *active* killing. In order to capture standard Western thinking, I insist, we need to introduce another distinction as well, the one between action where the death is intended and where it is merely foreseen, the doctrine of double effect. According to this principle it is always wrong intentionally to kill a patient, but it may be right to provide aggressive palliative care, with the intention of relieving pain, even if it can be foreseen that the patient will die from the care in question.

What about the principle of double effect, then, does *it* make sense? It does and it is clear, at least, that some standard rejections of it fail. Take the following famous example, given by Jonathan Glover and quoted in Chapter 9 of this anthology by Helga Kuhse:.

> When we are on a desert journey and I knowingly use all the drinking water for washing my shirts, my act may be described as one of 'washing shirts', or 'keeping up standards even in the desert', and our being out of water may be thought of as a [merely] foreseen [but not directly intended] consequence.
>
> But it is at least equally acceptable to include the consequence in the description of the act, which may then be described as one of 'using up the last of the water' or of 'putting our lives at risk'.[4]

Those who adhere to the doctrine of double effect would undoubtedly say that, irrespective of whether we describe the action as 'using up the last of the water', or as 'putting our lives at risk', crucial to its normative status is whether the intention is to kill the members of the expedition or not. Suppose it is not, then the intention is not forbidden, as such. We are not confronted with a case of intentional killing.

However, this does not mean that the action is right. We must also assess the importance of the intended first effect (keeping up standards, say), and compare it to the merely foreseen bad effect of the action (the death of the members of the expedition). Unless there is a reasonable proportion between these two effects, the action is not permissible. And it must he hard to make a case for the claim that the requirement of proportionality is fulfilled in the given example.

I see no need to go any deeper into the discussion about the doctrine of double effect. Those who adhere to the Sanctity-of-Life Doctrine probably agree with my assessment of this principle. They agree that it makes sense. According to this principle it is always wrong intentionally to kill a patient, but it may be right to provide aggressive palliative care with the intention of relieving pain, even if it can be foreseen that the patient will die from the care in question. This may be right provided that, in the circumstances, it is a good thing to have the patient free of pain, and, provided there is some reasonable proportion between the (first) good effect (the patient being free of pain) and the (second) bad effect (death being somewhat hastened).

Now, even if the principle of double effect is reasonably clear and comprehensible, in contemporary Western medicine we do not abide by it either. In most Western countries intentional killing of patients is legally tolerated. This is a claim I made in Chapter 2 of this book. Perhaps this is not explicitly acknowledged by medical doctors in general, but there is no denying that intentional killing of patients takes place and is tolerated, also outside Holland and Belgium. The much publicised Bland case I referred to in Chapter 3 bears witness to this.[5] In this case it was decided that a patient in a persistent vegetative state should not be artificially fed or hydrated any more. It is *obvious* that the intention behind the action (of not feeding or hydrating the patient) was to hasten death, I claimed.

Some doctors may, as Dan Brock points out, hesitate to admit this. They say that, when they stop feeding and hydrating the patient, they only honour the will (or presumed will) of the patient.

In many cases, this may certainly be correct. However, this does not mean that these doctors have not made the intention of the patient's their own. They have *two* intentions, first of all to do as the patient has informed them to do but also, for this very reason, to hasten the death of the patient. To see this we need only consider a parallel case outside medicine. What if a hired gun were to argue that he did not intend the death of his victim, he only wanted to fulfil a certain contract?

We would certainly not accept this argument for an answer. Moreover, to return to the medical setting, in the Bland case the wish of the patient was not known. This was certainly *passive* killing, but it is equally clear that it was *intentional*. Similar cases are easily found in most other Western countries. And they are all legally tolerated.

Does all this mean that neither the active/passive distinction nor the principle of double effect plays any role in Western medicine, and that the avenue to euthanasia is open? No, this is clearly not the case. However, only a *combination* of the two principles (of acts and omissions and the double effect) can substantiate the traditional approach. This was how I represented the situation in Chapter 2, putting the situation in Holland and Belgium to one side.

KILLING	Death intended	Death merely foreseen
Active	FORBIDDEN	TOLERATED
Passive	TOLERATED	TOLERATED

Figure

4. THE SANCTITY OF LIFE DOCTRINE

I am certainly no expert on what has come to be called the Sanctity-of-Life Doctrine. However, if we want to give it the most charitable interpretation, we should allow that it makes use of both the active/passive distinction and the principle of double effect, in the manner indicated above. In that case, the doctrine will not be at variance with standard medical practice in most Western countries, while it remains at variance with systems of euthanasia such as those practised in the Netherlands and Belgium and, as we shall see, with systems of physician-assisted suicide. The adherents of the doctrine, I suppose, should see both these features as merits of the doctrine. However, is this interpretation consistent with authoritative statements of it?

It is not, according to Helga Kuhse, who points to the Vatican's 1980 *Declaration on Euthanasia*, here reprinted in the appendix, where it is stated that 'euthanasia' or 'mercy-killing' is 'an action or an omission which of itself or by intention causes death.' [6]. Doesn't this mean that even passive killing, if intentional, is condemned by the Sanctity-of-Life Doctrine? For the doctrine certainly condemns all kinds of euthanasia.

First of all we should note that this passage of the declaration seems to rule out even palliative care, which, as a second and unintended effect, kills the patient. However, this is not how the declaration is usually understood. If the killing is active (the injection of morphine, for example), but the death of the patient merely foreseen, then even active killing may be permissible (if the proportionality requirement is satisfied as well). [7]

Note also that, typically, those who adhere to the Sanctity-of-Life Doctrine tend to agree that it is permissible to withhold or withdraw so-called 'extraordinary' or 'disproportionate' or 'burdensome' life-sustaining treatments. It is said indeed that when 'inevitable death is imminent in spite of the means used, it is permitted in conscience to take the decision to refuse forms of treatment that would only secure a precarious and burdensome prolongation of life ...'[9]

But here the intention does indeed seem to be to hasten death (or cutting shorter the period of dying). And the reason why this may be permissible, then, is that the measures taken are merely passive, or so I conjecture. How else could the Pope come to endorse this kind of treatment (or lack of treatment)?

What are we to do with the claim that the prohibition against killing includes, not only actions, but omissions as well? The most reasonable interpretation is this: only those omissions, that form part of active killing of patients, are prohibited *as such.* I think of cases like the one I mentioned in Chapter 2: You start to give an infusion, note that it should be slowed down, but omits to do anything about this. As a consequence the patient dies. According to my intuitions this is a clear case of *active* killing.

All this does not mean that there is explicit textual support in the Vatican's declaration for, say, the cessation of tube-feeding a person who is in a persistent vegetative state. Here it would be wrong in many cases to say that death is 'imminent'. And it is controversial, of course, whether the provision of nourishment and fluids through a tube is 'treatment'.

However, once the first step in this direction is taken (we need not secure a precarious and burdensome prolongation of life), then I see to *principled* reasons not to include these cases as well — while sticking to a principled rejection of the deliberate and active killing of innocent human beings. To do so would be to opt for the most plausible version of the Sanctity-of-Life Doctrine.

Or, to give an even more mundane example: Many people end their lives by ceasing to eat and drink. They feel that their lives are coming to an end, they have done what they could and should to 'conclude' them in a proper fashion, and they see no reason to keep as long as possible the last phase of their lives. Even eating and drinking have become burdensome (if not 'extraordinary', so at least 'disproportionate', they may argue) so this is no longer mandatory. Intentionally but passively, by ceasing to drink and eat, these people hasten their death.

Few people would feel that there is anything wrong in doing so. If I am right in my suggestion as to how the Sanctity-of-Life Doctrine should best be understood, then the adherents of this doctrine may safely concur in this judgement, while retaining their opposition to systems of active and intentional euthanasia as well as physician assisted suicide.

These people, who want to hold on to the Sanctity-of-Life Doctrine in the weak version here outlined, may take comfort in a recent study. According to this study, patients who chose to stop eating and drinking because they felt they were ready to die, who saw continued existence as pointless, and considered their quality of life poor in general did experience a good death. The survey was based on questions to nurses who had cared for such patients, and it showed that 85 percent of the patients died within 15 days after stopping food and fluids. On a scale from 0 (a very bad death) to 9 (a very good death), the median score for the quality of these deaths, as rated by the nurses, was 8.[10]

Doesn't this show, these people may argue, that there is no need to have a system of physician-assisted (active) suicide, when, passively, people who want to do so, can seek a morally legitimate and peaceful death by stopping to eat and drink? Or, to connect to the main theme of this book, they may insist that these people seek terminal sedation as a substitute for euthanasia.

Those who want to hold on to a strong version of the Sanctity-of-Life Doctrine, on the other hand, must condemn these patients who passively seek death by stopping eating and drinking or resort to terminal sedation just as hard as they condemn patients and doctors involved in physician assisted suicide or euthanasia.

Does not this constitute a good reason to opt for the weaker interpretation? I note that even a sceptic of the active/passive distinction and the double effect, such as Dr. Timothy Quill, find that some of his patients, their families and clinicians seem to hold on to it. Terminal sedation and voluntary refusal of hydration and nutrition are two options within palliative care, he writes, that '… provide a means of response for patients, families, and clinicians who oppose physician-assisted suicide.'[11]

5. LIVING HIGH AND LETTING DIE — AN ASIDE

This is not the place to undertake a complete examination of the Sanctity-of-Life Doctrine in all its applications. However, allow me to add just one more observation counting in favour of including a reference to the active/passive distinction in the doctrine. Unless we do, the doctrine seems to have far too radical implications with respect to aid. Unless we rely on the active/passive distinction we will have to say that people in the rich world, who are living high and letting (poor people in the poor part of the world) die, are flouting the doctrine. Now, being myself a utilitarian, I do not find this implication disturbing as such, but I doubt that the adherents of the doctrine would welcome it. Yet, it is impossible truthfully to deny that, by going on with our life projects in the rich part of the world, we allow that poor people die.

It might be tempting to say that, even if this is so, the deaths of these people are something we merely foresee, their deaths are not intended. Now, this may well be true, but even if it is, it doesn't let the adherent of the strong version of the Sanctity-of-Life Doctrine off the hook. For, certainly, in order to get off the hook they must not only show that the deaths of the people they allow to die are merely foreseen, they must also show that the requirement of proportionality is satisfied. And in relation to most trivial pursuits undertaken by rich people, the requirement of proportionality is not fulfilled.

It seems then that, unless we incorporate the active/passive distinction into the Sanctity-of-Life Doctrine, it will have as radical implications as utilitarianism with respect to foreign aid. But this is not a charitable interpretation of the doctrine.

This is not to say that adherents of the Sanctity-of-Life Doctrine must be complacent with respect to people in need. It is only to say that the doctrine as such has nothing to say about this. But the fact that the doctrine is silent on the question of aid means that it can be consistently combined will all sorts of principles about aid. The doctrine can be combined with a principle urging us to help people in need which is just as strong as, or even stronger than, utilitarianism. It can equally well be combined with a more commonsensical view on aid according to which we ought each to contribute a share that would be sufficient, if other people were to follow suit (irrespective of whether they do or not). Finally, it could be combined with the moral rights theory of the kind developed by Robert Nozick[12] according to which, unless we have caused their misery by violating their rights, we have no obligation to help people in distress.

6. CONCLUSION

I have argued that if we conceive of it in relational terms, the active/passive distinction makes sense. We can distinguish, in many cases, between active and passive killing. Moreover, I have argued that the doctrine of double effect makes sense as well. Again, we can distinguish in many cases between killing where death is intended, and cases where death is merely foreseen. However, according to standard Western medical ethical thinking, neither the active/passive distinction nor the doctrine of double effect draws, as such, a distinction between permissible and impermissible killing. If we want to account for standard medical ethical thinking in countries where euthanasia is prohibited, we need to have recourse to *both* the active/passive distinction and the doctrine of double effect. What is prohibited, in principle, is only intentional and active killing of patients.

Where does this leave the Sanctity-of-Life Doctrine? Is it at variance with standard medical ethical thinking, claiming that all instances of intentional killing are wrong? Or does it too rely on the active/passive distinction, condemning, in principle, only active and intentional killing?

It might be only natural for a thinker like Helga Kuhse, who wants to denounce the Sanctity-of-Life Doctrine, to give to it the strongest possible interpretation. But it is certainly true that even some thinkers, who like to defend the doctrine, such as Luke Gormally, cling to a strong interpretation of it. However, I have argued that, in the most *plausible* interpretation of the Sanctity-of-Life Doctrine, to wit, the one that makes it as attractive as possible, it *does* incorporate the active/passive distinction. This means that the doctrine is compatible with, and explains, standard medical ethical thinking. In such an interpretation the Sanctity-of-Life Doctrine allows both the active killing (in some cases) of patients, where death is merely foreseen, and the intentional killing of patients (in some cases), where the killing is passive (these are cases of allowing death to come). This means that the Sanctity-of-Life Doctrine can accept and explain standard Western medical ethical thinking. The doctrine can accept terminal sedation, as defined in this anthology. It can also accept the cessation of the provision of fluids and nourishment to patients in a persistent vegetative state. It can accept that dying patients, in order not to prolong their dying process (i.e., in order hasten death), voluntarily stop eating and drinking. And while doing so, the doctrine concurs in the widespread condemnation of systems of euthanasia such as the ones practised in the Netherlands and Belgium, as well as of systems of physician-assisted suicide, such as the one practised in Oregon. Moreover, in this interpretation the doctrine gives the result that, by living high and letting die, rich people in Western countries may well act immorally (they may flout some duty to assist people in distress), but this is not any immorality that stems from any violation of the Sanctity-of Life Doctrine as such. On this understanding of the Sanctity-of-Life Doctrine, what is inherently wrong and at variance with the doctrine is (only) the deliberate and active killing of innocent human beings.

Even if, personally, I do not adhere to the Sanctity-of-Life Doctrine, I have to admit that these are aspects of the doctrine that could probably persuade many people to accept it.

Torbjörn Tännsjö
Department of Philosophy
Stockholm University

NOTES

[1] See my *Hedonistic Utilitarianism* (Edinburgh: Edinburgh University Press, 1998), Chapter 2 about this.
[2] Dan W. Brock, *Life and Death: Philosophical Essays in Biomedical Ethics* (Cambridge: Cambridge University Press, 1993) 209. The passage is quoted by Brock in Chapter 7 of this book, p. 72.
[3] See p. 61 of this book.
[4] Jonathan Glover: *Causing Death and Saving Lives*, Harmondworth: Penguin Books, 1977, p. 90.
[5] Airedale NHS *Trust v. Bland* (1993).
[6] Sacred Congregation for the Doctrine of the Faith: *Declaration on Euthanasia*, Vatican City, 1980, p. 6; reprinted in the index of this book, p. 136.
[7] Sacred Congregation for the Doctrine of the Faith, op. cit., pp. 8-9; reprinted in the index of this book, p. 138..
[8] Sacred Congregation for the Doctrine of the Faith, op. cit., p. 10; reprinted in the index of this book, p. 139.
[9] Linda Gannzini et al., 'Nurses' Experience with Hospice Patients Who Refuse Food and Fluids to Hasten Death', *The New England Journal of Medicine,* Vol. 349, July 24, 2003, No. 4, pp. 359-365.
[10] Quill, T.E.l, and Boyck, I.R. 'Responding to Intractable Terminal Suffering: The Role of Terminal Sedation and Voluntary Refusal of Food and Fluids', *Ann Intern Med.* 2000;132:408-414.
[11] See his in *Anarchy, State, and Utopia* (Oxford: Blackwell, 1974).

NOTES

[illegible]

APPENDIX

AMERICAN ASSOCIATION FOR HOSPICE AND PALLIATIVE MEDICINE

POSITION STATEMENT ON SEDATION AT THE END-OF-LIFE

The American Academy of Hospice and Palliative Medicine regards sedation as a valid, ethically sound, and effective modality for relieving symptoms and suffering in some patients reaching the ends of their lives. Sedation is reserved for those in whom suffering is refractory and is not ameliorated by the application of other appropriate palliative care measures. These appropriate palliative care measures may include but are not limited to pain relief, non-pain symptom management, mental health care and spiritual counseling.

As one of the many forms of palliative treatment, the use of sedating medications is intended to decrease a patient's level of consciousness to mitigate the experience of suffering, but not to hasten the end of his or her life. Since this represents both an intention and an outcome that is beneficial, sedation in these cases is ethically justified. It is not analogous to euthanasia or physician-assisted suicide in which the primary intent is the death of the patient.

Careful patient and family assessment is a critical part of the evaluation process before sedation is provided for any individual. Hospice and palliative care programs and healthcare providers should undertake a thorough assessment of the alternative treatments available and engage in an open interdisciplinary decision-making

Torbjörn Tännsjö (ed.), Terminal Sedation: Euthanasia in Disguise?, 127-140.

process. This process must include the patient, if able, and family or other appropriate surrogate decision- makers, if the patient lacks decision-making capacity, in order to assure informed consent. Rarely, in emergent conditions, sedation may need to be applied prior to the appropriate discussions. In these cases, the discussions should take place as soon as possible once the comfort of the patient has been ensured.

Patients for whom sedation may be appropriate are most often near death as a result of an underlying disease process. Although the withdrawal of artificial hydration and nutrition commonly accompanies sedation, the decision to provide, withdraw, or withhold such treatments is separate from the decision whether or not to provide sedation.

Approved: AAHPM Ethics SIG August 31, 2002 AAHPM Board of Directors September 14, 2002

THE HOSPICE AND PALLIATIVE NURSES ASSOCIATION (HPNA)

(PITTSBURGH)

POSITION STATEMENT

The Hospice and Palliative Nurses Association (HPNA) is committed to compassionate care of persons at end of life. It is the position of the HPNA Board of Directors to:

Affirm the value of end of life care that includes aggressive and comprehensive symptom management.

Affirm the use of palliative sedation to manage refractory and unendurable symptoms in imminently dying patients as one method of aggressive and comprehensive symptom management.

Assert that hospice and palliative care nurses must possess sufficient knowledge about the issues surrounding the use of palliative sedation to inform patients, families, and other health care providers in making decisions about its use.

Direct those nurses who choose not to care for patient's receiving palliative sedation to continue to provide care until responsibility for care is transferred to an equally competent colleague.

Honor nurses rights to transfer care.

Affirm that consultation with, psychiatry, ethicists, chaplains, social workers, pharmacists and palliative care specialists may be needed to assure appropriateness of palliative sedation.

Oppose active euthanasia and assisted suicide as a means to relieve suffering.

Definition of Terms

Autonomy: a multidimensional ethical concept. It is the right of a capable person to decide his/her own course of action. Self-determination is a legal right.

Beneficence: an ethical duty to do well. It relates to promoting well being.

Dignity or Respect for person: a fundamental ethical principle. Dignity is the quality or state or being honored or valued. Respecting the body, values, beliefs, goals, privacy, actions and priorities of an autonomous adult preserves their dignity. This is a broader concept than autonomy.

Double Effect: an ethical principle based in Roman Catholic theology that is used when it is not possible to avoid all harmful effects. For an action to be ethically permissible under the principle of double effect, the act itself must be good or at least neutral (e.g. administering analgesic or sedative medication); the intention of the act is to produce a good effect (i.e., relief of suffering) even though a harmful effect (i.e. death) is foreseeable in some circumstances; the harmful effect of the act must not be the means to the good effect (i.e., death is not the means to relief of suffering); the good effect must outweigh or balance the harmful effect

Fidelity: the ethical imperative to keep promises. For health care providers, fidelity includes the promise not to abandon the patient.

Informed consent: a tenet of autonomy. To make an informed, autonomous decision, the person must have the capacity to understand the consequences of the decision; sufficient information about the treatment, desired outcomes, foreseeable consequences; intent to make the decision without coercion.

Imminent death: refers to death that is expected to occur within hours to days based on the person's current condition, progression of disease and symptom constellation.

Intent: the purpose or state of mind at the time of an action. Intent of the patient/ proxy and health care providers is a critical issue in ethical decision making around palliative sedation. Relief of suffering, not hastening or causing death, is the intent of palliative sedation.

Proxy decision making: allowed if the person lacks capacity to make an informed choice. Written advanced directives, substituted judgment based on subjective knowledge of the person's values, views on quality of life, goals, or the "best interest" of the person whose wishes and values are unknown based on benefits/burden weighing of recommended actions are the basis of such surrogate decisions.

Non-maleficence: the ethical duty to do no harm. When beneficence conflicts with non-maleficence, there is a greater duty to avoid inflicting harm.

Palliative Sedation: the monitored use of medications intended to provide relief of refractory symptoms by inducing varying degrees of unconsciousness but not death in terminally ill patients.

Refractory symptom: one that cannot be adequately controlled in a tolerable time frame despite aggressive use of usual therapies, and seems unlikely to be adequately controlled by further invasive or noninvasive therapies without excessive or intolerable acute or chronic side effects/complications.

Respite Sedation: the use of sedation for a brief, planned period to provide symptom relief and rest with the goal of returning to consciousness and pursuing future therapeutic and/or palliative therapies.

Suffering: a phenomenon of conscious existence. It is an aversive emotional experience brought on by an enduring perceived or actual threat to physical, psychological, social, spiritual well-being.

Withholding and withdrawing life sustaining therapy (LST): is legally and ethically permissible if it is the patient's fully informed and freely made wish; or such therapy is causing or will cause harm to the patient; or the therapy is not or will not benefit the patient in the future. Artificial hydration and nutrition may be withheld or withdrawn based on the same grounds.

Developed by:

Constance Dahlin, RN, CS, APN

Maureen Lynch, MS, RN, CS, AOCN, CHPN

Approved by the HPNA Board of Directors, June, 2003.

NORWEGIAN MEDICAL ASSOCIATION

GUIDELINES ON PALLIATIVE SEDATION

GUIDELINES

1. By palliative sedation of dying patients is meant a medically induced reduction of consciousness intended to alleviate suffering that cannot otherwise be alleviated.

2. Palliative sedation should only be resorted to in exceptional cases, where an intolerable suffering is caused by, and is dominated by, physical symptoms. Psychological suffering alone is not an indication for palliative sedation.

3. Palliative sedation of a dying patient should only be contemplated when the patient can be expected to have at most a few days more to live.

4. The causes of the pain felt by the patient must have been adequately diagnosed. All the alternative treatments of the individual symptoms must have been tried, or, at least, must have, upon serious consideration, been found to be useless. If the clinic lacks the necessary resources to help the patient, then the patient should be referred to, or the clinic should seek competent advice from, a clinic where the adequate resources to deal with pain control exists.

If it is obvious that a lack of resources is an obstacle to optimal treatment and care, unless sedation is resorted to, the responsible physician should report this to his chief of staff, or to the supervising authorities.

5. The chief physician at the clinic is responsible for the decision to initiate and to perform palliative sedation.

A decision to do so must be founded on an encompassing professional assessment of the entire situation of the patient, and it should be reached after consultations have been made with nursing personnel and with others doctors who know the patient, or who can contribute their competence.

6. The patient should, if he is capable of doing so, give his explicit consent to the treatment. The patient should be given information about his diagnosis and prognosis, what palliative sedation means (how deep the sedation will be and how

long it is supposed to go on), about risks associated with the treatment, and alternatives to it.

7. Unless the patient rejects this, his close ones should also be informed and engaged in the decision procedure. Close ones have a right to information, but should not be burdened with the decision as such to start the treatment.

8. The patient should only be sedated up to the point where the suffering has been alleviated.

9. Even if, in some cases, it is reasonable to assume that the sedation will go on until the patient is dead, arousal of the patient should be assessed and attempted. If, when the patient regains consciousness, it is clear that the situation of the patient is still unbearable, it may be defensible to sedate the patient once again, without any intention that the patient should regain consciousness.

10. The patient should be adequately supervised, in order to render possible control of his state of consciousness and to ascertain that the patient does not suffocate; the effects of the treatment should be evaluated. The supervision should include such possible side-effects of the treatment, that could be properly managed.

11. A patient who has stopped drinking will need no intravenous fluids. If the patient is still capable of drinking, so that the sedation as such has rendered the patient incapable of doing so, then intravenous fluids should be given to the patient. If an infusion has been started before the patient was sedated, it should be continued.

12. The treatment should be documented. The following aspects should be given weight:

— The rationale behind the sedation.
— The manner in which the decision to sedate was reached.
— Information to the patient and his close ones.
— The opinions with respect to the treatment of the patient and his close ones.
— How the sedation was implemented and supervised.

The guidelines were adopted in 2001.
(Unauthorised translation performed by the editor of this volume)

SACRED CONGREGATION FOR THE DOCTRINE OF THE FAITH

DECLARATION ON EUTHANASIA

1. INTRODUCTION

The rights and values pertaining to the human person occupy an important place among the questions discussed today. In this regard, the Second Vatican Ecumenical Council solemnly reaffirmed the lofty dignity of the human person, and in a special way his or her right to life. The Council therefore condemned crimes against life "such as any type of murder, genocide, abortion, euthanasia, or willful suicide" (Pastoral Constitution "Gaudium et spes," no. 27).

More recently, the Sacred Congregation for the Doctrine of the Faith has reminded all the faithful of Catholic teaching on procured abortion.[1] The Congregation now considers it opportune to set forth the Church's teaching on euthanasia.

It is indeed true that, in this sphere of teaching, the recent Popes have explained the principles, and these retain their full force,[2] but the progress of medical science in recent years has brought to the fore new aspects of the question of euthanasia, and these aspects call for further elucidation on the ethical level.

In modern society, in which even the fundamental values of human life are often called into question, cultural change exercises an influence upon the way of looking at suffering and death; moreover, medicine has increased its capacity to cure and to prolong life in particular circumstances, which sometimes give rise to moral problems. Thus people living in this situation experience no little anxiety about the meaning of advanced old age and death. They also begin to wonder whether they have the right to obtain for themselves or their fellowmen an "easy death," which would shorten suffering and which seems to them more in harmony with human dignity.

A number of Episcopal Conferences have raised questions on this subject with the Sacred Congregation for the Doctrine of the Faith. The Congregation, having sought the opinion of experts on the various aspects of euthanasia, now wishes to respond to the Bishops' questions with the present Declaration, in order to help them to give correct teaching to the faithful entrusted to their care, and to offer them elements for reflection that they can present to the civil authorities with regard to this very serious matter.

The considerations set forth in the present document concern in the first place all those who place their faith and hope in Christ, who, through His life, death and resurrection, has given a new meaning to existence and especially to the death of the Christian, as St. Paul says: "If we live, we live to the Lord, and if we die, we die to the Lord" (Rom. 14:8; cf. Phil. 1:20).

As for those who profess other religions, many will agree with us that faith in God the Creator, Provider and Lord of life--if they share this belief--confers a lofty dignity upon every human person and guarantees respect for him or her.

It is hoped that this Declaration will meet with the approval of many people of good will, who, philosophical or ideological differences notwithstanding, have nevertheless a lively awareness of the rights of the human person. These rights have often, in fact, been proclaimed in recent years through declarations issued by International Congresses,[3] and since it is a question here of fundamental rights inherent in every human person, it is obviously wrong to have recourse to arguments from political pluralism or religious freedom in order to deny the universal value of those rights.

2. THE VALUE OF HUMAN LIFE

Human life is the basis of all goods, and is the necessary source and condition of every human activity and of all society. Most people regard life as something sacred and hold that no one may dispose of it at will, but believers see in life some thing greater, namely, a gift of God's love, which they are called upon to preserve and make fruitful. And it is this latter consideration that gives rise to the following consequences:

1. No one can make an attempt on the life of an innocent person without opposing God's love for that person, without violating a fundamental right, and therefore without committing a crime of the utmost gravity.[4]
2. Everyone has the duty to lead his or her life in accordance with God's plan. That life is entrusted to the individual as a good that must bear fruit already here on earth, but that finds its full perfection only in eternal life.
3. Intentionally causing one's own death, or suicide, is therefore equally as wrong as murder; such an action on the part of a person is to be considered as a rejection of God's sovereignty and loving plan. Furthermore, suicide is also often a refusal of love for self, the denial of the natural instinct to live, a flight from the duties of justice and charity owed to one's neighbor, to various communities or to the whole of society--although, as is generally recognized, at times there are psychological factors present that can diminish responsibility or even completely remove it.

However, one must clearly distinguish suicide from that sacrifice of one's life whereby for a higher cause, such as God's glory, the salvation of souls or the service of one's brethren, a person offers his or her own life or puts it in danger (cf. Jn. 15:14).

3. EUTHANASIA

In order that the question of euthanasia can be properly dealt with, it is first necessary to define the words used. Etymologically speaking, in ancient times euthanasia meant an easy death without severe suffering. Today one no longer thinks of this original meaning of the word, but rather of some intervention of medicine whereby the suffering of sickness or of the final agony are reduced, sometimes also with the danger of suppressing life prematurely. Ultimately, the word euthanasia is used in a more particular sense to mean "mercy killing," for the purpose of putting an end to extreme suffering, or saving abnormal babies, the mentally ill or the incurably sick from the prolongation, perhaps for many years, of a miserable life, which could impose too heavy a burden on their families or on society.

It is, therefore, necessary to state clearly in what sense the word is used in the present document.

By euthanasia is understood an action or an omission which of itself or by intention causes death, in order that all suffering may in this way be eliminated. Euthanasia's terms of reference, therefore, are to be found in the intention of the will and in the methods used.

It is necessary to state firmly once more that nothing and no one can in any way permit the killing of an innocent human being, whether a fetus or an embryo, an infant or an adult, an old person, or one suffering from an incurable disease, or a person who is dying. Furthermore, no one is permitted to ask for this act of killing, either for himself or herself or for another person entrusted to his or her care, nor can he or she consent to it, either explicitly or implicitly. Nor can any authority legitimately recommend or permit such an action. For it is a question of the violation of the divine law, an offense against the dignity of the human person, a crime against life, and an attack on humanity.

It may happen that, by reason of prolonged and barely tolerable pain, for deeply personal or other reasons, people may be led to believe that they can legitimately ask for death or obtain it for others. Although in these cases the guilt of the individual may be reduced or completely absent, nevertheless the error of judgment into which the conscience falls, perhaps in good faith, does not change the nature of this act of killing, which will always be in itself something to be rejected. The pleas of gravely ill people who sometimes ask for death are not to be understood as implying a true desire for euthanasia; in fact, it is almost always a case of an anguished plea for help and love. What a sick person needs, besides medical care, is love, the human and supernatural warmth with which the sick person can and ought to be surrounded by all those close to him or her, parents and children, doctors and nurses.

4. THE MEANING OF SUFFERING FOR CHRISTIANS AND THE USE OF PAINKILLERS

Death does not always come in dramatic circumstances after barely tolerable sufferings. Nor do we have to think only of extreme cases. Numerous testimonies which confirm one another lead one to the conclusion that nature itself has made provision to render more bearable at the moment of death separations that would be terribly painful to a person in full health. Hence it is that a prolonged illness, advanced old age, or a state of loneliness or neglect can bring about psychological conditions that facilitate the acceptance of death.

Nevertheless the fact remains that death, often preceded or accompanied by severe and prolonged suffering, is something which naturally causes people anguish.

Physical suffering is certainly an unavoidable element of the human condition; on the biological level, it constitutes a warning of which no one denies the usefulness; but, since it affects the human psychological makeup, it often exceeds its own biological usefulness and so can become so severe as to cause the desire to remove it at any cost.

According to Christian teaching, however, suffering, especially suffering during the last moments of life, has a special place in God's saving plan; it is in fact a sharing in Christ's passion and a union with the redeeming sacrifice which He offered in obedience to the Father's will. Therefore, one must not be surprised if some Christians prefer to moderate their use of painkillers, in order to accept voluntarily at least a part of their sufferings and thus associate themselves in a conscious way with the sufferings of Christ crucified (cf. Mt. 27:34). Nevertheless it would be imprudent to impose a heroic way of acting as a general rule. On the contrary, human and Christian prudence suggest for the majority of sick people the use of medicines capable of alleviating or suppressing pain, even though these may cause as a secondary effect semi-consciousness and reduced lucidity. As for those who are not in a state to express themselves, one can reasonably presume that they wish to take these painkillers, and have them administered according to the doctor's advice.

But the intensive use of painkillers is not without difficulties, because the phenomenon of habituation generally makes it necessary to increase their dosage in order to maintain their efficacy. At this point it is fitting to recall a declaration by Pius XII, which retains its full force; in answer to a group of doctors who had put the question: "Is the suppression of pain and consciousness by the use of narcotics ... permitted by religion and morality to the doctor and the patient (even at the approach of death and if one foresees that the use of narcotics will shorten life)?" the Pope said: "If no other means exist, and if, in the given circumstances, this does not prevent the carrying out of other religious and moral duties: Yes."[5] In this case, of course, death is in no way intended or sought, even if the risk of it is reasonably taken; the intention is simply to relieve pain effectively, using for this purpose painkillers available to medicine.

However, painkillers that cause unconsciousness need special consideration. For a person not only has to be able to satisfy his or her moral duties and family obligations; he or she also has to prepare himself or herself with full consciousness for meeting Christ. Thus Pius XII warns: "It is not right to deprive the dying person of consciousness without a serious reason."[6]

5. DUE PROPORTION IN THE USE OF REMEDIES

Today it is very important to protect, at the moment of death, both the dignity of the human person and the Christian concept of life, against a technological attitude that threatens to become an abuse. Thus some people speak of a "right to die," which is an expression that does not mean the right to procure death either by one's own hand or by means of someone else, as one pleases, but rather the right to die peacefully with human and Christian dignity. From this point of view, the use of therapeutic means can sometimes pose problems.

In numerous cases, the complexity of the situation can be such as to cause doubts about the way ethical principles should be applied. In the final analysis, it pertains to the conscience either of the sick person, or of those qualified to speak in the sick person's name, or of the doctors, to decide, in the light of moral obligations and of the various aspects of the case.

Everyone has the duty to care for his or her own health or to seek such care from others. Those whose task it is to care for the sick must do so conscientiously and administer the remedies that seem necessary or useful.

However, is it necessary in all circumstances to have recourse to all possible remedies?

In the past, moralists replied that one is never obliged to use "extraordinary" means. This reply, which as a principle still holds good, is perhaps less clear today, by reason of the imprecision of the term and the rapid progress made in the treatment of sickness. Thus some people prefer to speak of "proportionate" and "disproportionate" means. In any case, it will be possible to make a correct judgment as to the means by studying the type of treatment to be used, its degree of complexity or risk, its cost and the possibilities of using it, and comparing these elements with the result that can be expected, taking into account the state of the sick person and his or her physical and moral resources.

In order to facilitate the application of these general principles, the following clarifications can be added:

— If there are no other sufficient remedies, it is permitted, with the patient's consent, to have recourse to the means provided by the most advanced medical techniques, even if these means are still at the experimental stage and are not without a certain risk. By accepting them, the patient can even show generosity in the service of humanity.

— It is also permitted, with the patient's consent, to interrupt these means, where the results fall short of expectations. But for such a decision to be made, account will have to be taken of the reasonable wishes of the patient and the patient's family, as also of the advice of the doctors who are specially competent in the matter. The latter may in particular judge that the investment in instruments and personnel is disproportionate to the results foreseen; they may also judge that the techniques applied impose on the patient strain or suffering out of proportion with the benefits which he or she may gain from such techniques.

— It is also permissible to make do with the normal means that medicine can offer. Therefore one cannot impose on anyone the obligation to have recourse to a technique which is already in use but which carries a risk or is burdensome. Such a refusal is not the equivalent of suicide; on the contrary, it should be considered as an acceptance of the human condition, or a wish to avoid the application of a medical procedure disproportionate to the results that can be expected, or a desire not to impose excessive expense on the family or the community.

— When inevitable death is imminent in spite of the means used, it is permitted in conscience to take the decision to refuse forms of treatment that would only secure a precarious and burdensome prolongation of life, so long as the normal care due to the sick person in similar cases is not interrupted. In such circumstances the doctor has no reason to reproach himself with failing to help the person in danger.

6. CONCLUSION

The norms contained in the present Declaration are inspired by a profound desire to serve people in accordance with the plan of the Creator. Life is a gift of God, and on the other hand death is unavoidable; it is necessary, therefore, that we, without in any way hastening the hour of death, should be able to accept it with full responsibility and dignity. It is true that death marks the end of our earthly existence, but at the same time it opens the door to immortal life. Therefore, all must prepare themselves for this event in the light of human values, and Christians even more so in the light of faith.

As for those who work in the medical profession, they ought to neglect no means of making all their skill available to the sick and the dying; but they should also remember how much more necessary it is to provide them with the comfort of boundless kindness and heartfelt charity. Such service to people is also service to Christ the Lord, who said: "As you did it to one of the least of these my brethren, you did it to me" (Mt. 25:40).

At the audience granted to the undersigned Prefect, His Holiness Pope John Paul II approved this Declaration, adopted at the ordinary meeting of the Sacred Congregation for the Doctrine of the Faith, and ordered its publication.

Rome, the Sacred Congregation for the Doctrine of the Faith, May 5, 1980.

Franjo Cardinal Seper Prefect

+ Jerome Hamer, O.P. Tit. Archbishop of Lorium Secretary

NOTES

[1] "Declaration on Procured Abortion," November 18, 1974: AAS 66 (1974), pp 730-747.
[2] Pius XII, "Address to those attending the Congress of the International Union of Catholic Women's Leagues," September 11, 1947: AAS 39 (1947), p. 483; "Address to the Italian Catholic Union of Midwives," October 29, 1951: AAS 43 (1951), pp. 835- 854; "Speech to the members of the International Office of Military Medicine Documentation," October 19, 1953: AAS 45 (1953), pp. 744-754; "Address to those taking part in the IXth Congress of the Italian Anaesthesiological Society," February 24, 1957: AAS 49 (1957). p. 146; cf. also "Address on reanimation," November 24, 1957: AAS 49 (1957), pp. 1027-1033; Paul VI, "Address to the members of the United Nations Special Committee on Apartheid," May 22, 1974: AAS 66 (1974), p. 346; John Paul II: "Address to the bishops of the United States of America," October 5, 1979: AAS 71 (1979), p. 1225.
[3] One thinks especially of Recommendation 779 (1976) on the rights of the sick and dying, of the Parliamentary Assembly of the Council of Europe at its XXVIIth Ordinary Session; cf. Sipeca, no. 1, March 1977, pp. 14-15.
[4] We leave aside completely the problems of the death penalty and of war, which involve specific considerations that do not concern the present subject.
[5] Pius XII, "Address" of February 24, 1957: AAS 49 (1957), p. 147.
[6] Pius XII, ibid., p. 145; cf. "Address" of September 9, 1958: AAS 50 (1958), p. 694.

TERMINAL SEDATION 1990-2003

BIBLIOGRAPHY

Ashby M.J., Stoffell B. Artificial hydration and alimentation at the end of life: a reply to Craig. *J Med Ethics*. 1995 Jun;21(3):135-40.

Ashby M.J., The fallacies of death causation in palliative care. *Medical Journal of Australia* 1997;166:176.

Ashby, M.J. (1998) Palliative care, death causation, public policy and the law. *Progress in Palliative Care* 6(3), 69-77.

Ashby M.J. On causing death. *Medical Journal of Australia* 2001;175:517-518.

Billings JA, Block SD, 'Slow Euthanasia', *Journal of Palliaative Care* 1996;12(4):21-30.

Brody, H., 'Commentary on Billings and Block's "Slow Euthanasia"', *JPalliat Care,* 1996;12:38-41.

Broekaert B Palliative sedation defined or why and when terminal sedation is not euthanasia. Abstract 1st Congress RDPC, December 2000, Berlin (Germany) *Journal of Pain and Symptom Management* 20 (6), 2000, S58.

Broeckaert B, Nunex-Olarte J.M, 'Sedation in palliative care: facts and concepts'. In: ten Have H & D Clark (ed.) *The ethics of palliative care: European perspectives*. Buckingham. Open University Press, 2002. 166-180.

Broeckaert B., 'Palliative sedation: ethical aspects'. In: C Gastmans (ed.) *Between technology and humanity, the impact of technology on health care ethics*. Leuven University Press, 2002, 239-255.

Chater S, Viola R, Paterson J, Jarvis V. Sedation for intractable distress in the dying--a survey of experts. *Palliat Med*. 1998 Jul;12(4):255-69.

Cherny NI, Portenoy RK. Sedation in the management of refractory symptoms: guidelines for evaluation and treatment. *J Palliat Care*. 1994 Summer;10(2):31-8.

Chiu TY, Hu WY, Cheng SY, Chen CY. Ethical dilemmas in palliative care: a study in Taiwan. *J Med Ethics*. 2000 Oct;26(5):353-7.

Chiu TY, Hu WY, Lue BH, Cheng SY, Chen CY. Sedation for refractory symptoms of terminal cancer patients in Taiwan. *J Pain Symptom Manage*. 2001 Jun;21(6):467-72.

Cleeland, C.S., Gonin, R., Hatfield, A., Edmonson, J.H., Blum, R.H., Stewart, J.A. and et al. Pain and its treatment in outpatients with metastatic cancer. *N Engl J Med* 1994;330, 592-596.

Collins J. Case presentation: terminal sedation in a pediatric patient. *J Pain Symptom Manage*. 1998 Apr;15(4):258-9.

Cowan JD, Walsh D. Terminal sedation in palliative medicine--definition and review of the literature. *Support Care Cancer*. 2001 Sep;9(6):403-7.

Cowan JD, Palmer TW. Practical guide to palliative sedation. *Curr Oncol Rep*. 2002 May;4(3):242-9.

Coyle N, Truog RD. HealthCare Ethics Forum '94: pain management and sedation in the terminally ill. *AACN Clin Issues Crit Care Nurs*. 1994 Aug;5(3):360-5.

Craig, G. M. 'On withholding nutrition and hydration in the terminally ill: has palliative medicine gone too far?', *Journal of medical ethics*, Vol. 20, 1994: 139-143.

Craig G. Is sedation without hydration or nourishment in terminal care lawful? *Med Leg J*. 1994;62 (Pt 4):198-201.Erratum in: *Med Leg* J 1995;63(Pt 1):31.

Craig G. Palliative care from the perspective of a consultant geriatrician: the dangers of withholding hydration. *Ethics and Medicine* 1999;15(1):15-19.

Craig G., Terminal Sedation. *Catholic Medical Quarterly* 2002;52(1):14-17.

Torbjörn Tännsjö (ed.), Terminal Sedation: Euthanasia in Disguise?, 141-144.

Cranford RE, Gensinger R. Hospital policy on terminal sedation and euthanasia. *HEC Forum.* 2002;14:259-64.

Davidoff F. Publication of Papers on Assisted Suicide and Terminal Sedation. *Ann Intern Med.* 2000 Oct 3;133(7):566.

Devettere RJ. Sedation before ventilator withdrawal: can it be justified by double effect and called "allowing a patient to die" *J Clin Ethics.* 1991 Summer;2(2):122-4.

Ducharme, H.M., 'Thrift-Euthanasia, In Theory and In Practice' *Law and Medicine: Current Legal Issues* 2000, Volume 3, edited by Michael Freeman and Andrew D.E. Lewis (Oxford University Press, 2000), pp. 493-525.

Ducharme H.M., Total sedation as existential euthanasia; www.cbhd.org/resources/aps/ducharme-total-_sedation.htm

Dunlop R.J., Ellershaw J.E., Baines M.J., Sykes N., and Saunders C.M., 'On withholding nutrition and hydration in the terminally ill: has palliative medicine gone too far? A reply', *Journal of medical ethics*, Vol. 21, 1995: 141-143.

Enck RE. Drug-induced terminal sedation for symptom control. *Am J Hosp Palliat Care.* 1991 Sep-Oct;8(5):3-5.

Enck RE. Terminal sedation. *Am J Hosp Palliat Care.* 2000 May-Jun;17(3):148-9.

Fainsinger R, Miller MJ, Bruera E, Hanson J, Maceachern T. Symptom control during the last week of life on a palliative care unit. *J Palliat Care.* 1991 Spring;7(1):5-11.

Fainsinger R.L., Tapper M., Bruera E., A perspective on the management of delirium in terminally ill patients on a palliative care unit. *J Palliat Care* 1993;9(3):4-8.

Fainsinger R.L., Use of sedation by a hospital palliative care support team. *J Palliat Care* 1998;14(1):51-54.

Fainsinger RL, Waller A, Bercovici M, Bengtson K, Landman W, Hosking M, Nunez-Olarte JM, deMoissac D. A multicentre international study of sedation for uncontrolled symptoms in terminally ill patients. *Palliat Med.* 2000 Jul;14(4):257-65.

Fainsinger RL, De Moissac D, Mancini I, Oneschuk D. Sedation for delirium and other symptoms in terminally ill patients in Edmonton. *J Palliat Care.* 2000 Summer;16(2):5-10.

Fine, P., MD, "Total Sedation in End-of-Life Care: Clinical Considerations", *Journal of Hospice and Palliative Nursing*, Vol. 3, No. 3, July-September 2001.

Fleischman A. Commentary: ethical issues in pediatric pain management and terminal sedation. *J Pain Symptom Manage.* 1998 Apr;15(4):260-1.

Gauthier CC. Active voluntary euthanasia, terminal sedation, and assisted suicide. *J Clin Ethics.* 2001 Spring;12(1):43-50.

Gormally L., Palliative treatment and ordinary care. In: J. Vial Correa & E Sgreccia (eds) *The Dignity of the Dying Person.* Vatican City: Libreria Editrice Vaticana 2000: 252-266.

Greene, W.R. and Davis, W.H. Titraded intravenous barbiturates in the control fo symptoms in patients with terminal cancer. *South Med J* 1991;84:332-337.

Hallenbech JL, 'Terminal Sedation for Intractable Distress, Not Slow Euthanasia but Proper Responsiveness to Suffering' *West. J. Med.* 1999; 171: 222-223.

Hardy J. Sedation in terminally ill patients. *Lancet.* 2000 Dec 2;356(9245):1866-7.

Heyse-Moore L. Related Articles, 'Terminal restlessness and sedation: a note of caution', *Palliat Med.* 2003;17:469.

Jackson WC. Palliative sedation vs. terminal sedation: what's in a name*? Am J Hosp Palliat Care.* 2002 Mar-Apr;19(2):81-2.

Jansen LA, Sulmasy DP. Sedation, alimentation, hydration, and equivocation: careful conversation about care at the end of life. *Ann Intern Med.* 2002 Jun 4;136(11):845-9.

Kenny NP, Frager G. Refractory symptoms and terminal sedation of children: ethical issues and practical management. *J Palliat Care.* 1996 Autumn;12(3):40-5.

Kingsbury, R.J., 'Palliative Sedation: May We Sleep Before We Die?', *Dignity newsletter*, Summer, 2001.

Krakauer EL, Penson RT, Truog RD, King LA, Chabner BA, Lynch TJ Jr. Sedation for intractable distress of a dying patient: acute palliative care and the principle of double effect. *Oncologist.* 2000;5(1):53-62.

Lanuke K, Fainsinger RL, DeMoissac D, Archibald J. 'Two remarkable dyspneic men: when should terminal sedation be administered?', *J Palliat Med.* 2003;6:277-281.

Loewy EH. Terminal sedation, self-starvation, and orchestrating the end of life. *Arch Intern Med.* 2001 Feb 12;161(3):329-32.

Lynn J. Terminal sedation. *N Engl J Med.* 1998 Apr 23;338(17):1230; author reply 1230-1.

Materstvedt LJ, Kaasa S. Euthanasia and physician-assisted suicide in Scandinavia--with a conceptual suggestion regarding international research in relation to the phenomena. *Palliat Med.* 2002 Jan;16(1):17-32.

Morita T, Tsunoda J, Inoue S, Chihara S. Terminal sedation for existential distress. *Am J Hosp Palliat Care.* 2000 May-Jun;17(3):189-95.

Morita T, Tsuneto S, Shima Y. Proposed definitions for terminal sedation. *Lancet.* 2001 Jul 28;358(9278):335-6.

Morita T, Tsuneto S, Shima Y.Definition of sedation for symptom relief: a systematic literature review and a proposal of operational criteria. *J Pain Symptom Manage.* 2002;24:447-53.

Morita T, Hirai K, Okazaki Y. Preferences for palliative sedation therapy in the Japanese general population.*J Palliat Med.* 2002;5:375-85.

Morita T, Akechi T, Sugawara Y, Chihara S, Uchitomi Y. Practices and attitudes of Japanese oncologists and palliative care physicians concerning terminal sedation: a nationwide survey. *J Clin Oncol.* 2002 Feb 1;20(3):758-64.

Morita T, Hirai K, Akechi T, Uchitomi Y., 'Similarity and difference among standard medical care, palliative sedation therapy, and euthanasia: a multidimensional scaling analysis on physicians' and the general population's opinions'. *J Pain Symptom Manage.* 2003;25:357-62.

Mount B, 'Morphine drips, terminal sedation, and slow euthanasia: definitions and facts, not anecdotes'. *J Palliat Care.* 1996;12:31-37.

Muller-Busch HC, Andres I, Jehser T., 'Sedation in palliative care - a critical analysis of 7 years experience', *BMC Palliat Care.* 2003;2:2.

Norris, P. Palliative Care and Killing: Understanding Ethical Distinctions. *Bioethics-Forum* 1998;7(4):382-387.

Nunez Olarte JM, Guillen DG. Cultural issues and ethical dilemmas in palliative and end-of-life care in Spain. *Cancer Control.* 2001 Jan-Feb;8(1):46-54.

O'Connor,M. Kissane D.W., and Spruyt O., 'Sedation of the terminally ill – a clinical perspective', *Monash Bioethics Review*, Vol. 18, 1999, p. 22.

Orentlicher D. The Supreme Court and terminal sedation: rejecting assisted suicide, embracing euthanasia. *Hastings Constit Law Q.* 1997 Summer;24(4):947-68.

Orentlicher D. The alleged distinction between euthanasia and the withdrawal of life-sustaining treatment: conceptually incoherent and impossible to maintain. *Univ Ill Law Rev.* 1998;1998(3):837-59.

Powell T, Kornfeld DS. Commentary on "Sedation before ventilator withdrawal. *J Clin Ethics.* 1991 Summer;2(2):126-7.

Quill, T.E., Dresser, R. and Brock, D.W. The rule of double effect — a critique of its role in end-of-life decision making. *N Engl J Med* 1997;337:1768-1771.

Quill, T.E, Lo, B., Brock D.W., 'Palliative Options of Last Resort', *JAMA*, Vol. 278, 1997: 2099-2104.

Quill TE, Byock IR. Responding to intractable terminal suffering: the role of terminal sedation and voluntary refusal of food and fluids. ACP-ASIM End-of-Life Care Consensus Panel. American College of Physicians-American Society of Internal Medicine. *Ann Intern Med.* 2000 Mar 7;132(5):408-14. Review. Erratum in: *Ann Intern Med* 2000 Jun 20;132(12):1011.

Quill TE, Lee BC, Nunn S. Palliative treatments of last resort: choosing the least harmful alternative. University of Pennsylvania Center for Bioethics Assisted Suicide Consensus Panel. *Ann Intern Med.* 2000 Mar 21;132(6):488-93.

Radbruch L (2002) Reflections on the use of sedation in terminal care. *European Journal of Palliative Care,* 9(6): 237- 239.

Rosen EJ. Commentary: a case of "terminal sedation" in the family. *J Pain Symptom Manage.* 1998 Dec;16(6):406-7.

Rousseau P. Hospice and palliative care. *Dis Mon.* 1995 Dec;41(12):779-842. Erratum in: *Dis Mon* 1996 Sep;42(9):608.

Rousseau P. Commentary: Terminal sedation in the care of dying patients. *Arch Intern Med.* 1996 Sep 9;156(16):1785-1786.

Rousseau P. The ethical validity and clinical experience of palliative sedation. *Mayo Clin Proc.* 2000 Oct;75(10):1064-9.

Rousseau P. Existential suffering and palliative sedation: a brief commentary with a proposal for clinical guidelines. *Am J Hosp Palliat Care*. 2001 May-Jun;18(3):151-3.

Rousseau PC. Palliative sedation. *Am J Hosp Palliat Care*. 2002;19:295-7.

Roy, D. 'Need They Sleep Before They Die?' *J. Pall. Care* 1990; 6:(3) 3-4.

Snyder L. Publication of Papers on Assisted Suicide and Terminal Sedation. *Ann Intern Med*. 2000 Oct 3;133(7):565-566.

Schneiderman, L.J., 'Is it morally justifiable not to sedate this patient before ventilator withdrawal', *The Journal of Clinical Ethics*, Vol.2, 1991: 129-130.

Seymour, J.E., Clark, D., Gott M., Bellamy, G., and Ahmedzai, S. (2002). Good deaths, bad deaths: older people's assessments of risks and benefits in the use of morphine and terminal sedation in end of life care. *Health, Risk and Society* 4(3): 287-303.

Shaiova L. Case presentation: "terminal sedation" and existential distress. *J Pain Symptom Manage*. 1998 Dec;16(6):403-4.

Smith GP 2nd. Terminal sedation as palliative care: revalidating a right to a good death. *Camb Q Healthc Ethics*. 1998 Fall;7(4):382-7.

Stone P, Phillips C, Spruyt O, Waight C. A comparison of the use of sedatives in a hospital support team and in a hospice. *Palliat Med*. 1997 Mar;11(2):140-4.

Sulmasy, D.P. and Pellegrino, E.D. The role of double effect: clearing up the double talk. *Arch Intern Med* 1999;159:545-550.

Sulmasy DP, Ury WA, Ahronheim JC, Siegler M, Kass L, Lantos J, Burt RA, Foley K, Payne R, Gomez C, Krizek TJ, Pellegrino ED, Portenoy RK. Publication of papers on assisted suicide and terminal sedation. *Ann Intern Med*. 2000 Oct 3;133(7):564-6.

Sykes N. The management of difficult pain and other symptoms at the end of life. Pain in Europe III, advances in pain research and therapy. *Abstract* ORC/06,71-72.

Sykes N, Thorns A. 'Sedative use in the last week of life and the implications for end-of-life decision making. *Arch Intern Med*. 2003;163:341-4.

Sykes N and Thorns A (2003) 'The use of opioids and sedatives at the end of life' *The Lancet Oncology*, 4:312-318.

Tännsjö, T., 'Terminal Sedation – A Compromise in the Euthanasia Debate?', *Bulletin of Medical Ethics*, No. 163, November 2000, pp. 13–22.

Thorns A, Sykes N. 'Opioid use in last week of life and implications for end-of-life decision making'. *The Lancet*. 2000;356:398-9.

Tonelli MR. Terminal sedation. *N Engl J Med*. 1998 Apr 23;338(17):1230; author reply 1230-1.

Troug, D, Arnold, K.H., and Rockoff M.A., 'Sedation Before Ventilator Withdrawal: Medical and Ethical Considerations', *The Journal of Clinical Ethics*, Vol. 2, 1991:127.

Troug, RD, Berde CB, Mitchell C, Grier HE, 'Barbiturates in the care of the terminally ill. *N Engl J Med*. 1992;327:1678-1682.

Valko NG., 'Should sedation be terminal?', *Natl Cathol Bioeth Q*. 2002;2:601-8.

Wein S. Sedation in the imminently dying patient. *Oncology* (Huntingt). 2000 Apr;14(4):585-92; discussion 592, 597-8, 601.

Ventafridda V, Ripamonti C, De Conno F, Tamburini M, Cassileth BR. Symptom prevalence and control during cancer patients' last days of life. *J Palliat Care*. 1990 Autumn;6(3):7-11.

Williams G., The principle of double effect and terminal sedation. *Medical Law Review* 2001;9:41-53.

Voltz, R. and Borasio, G.D. Palliative therapy in the terminal stage of neurological disease. *J Neurol* 1997;244:2-10.

INDEX

A

B

D

E

O

P

Q

R

S

T

U

W